A Practical Guide to

Systems Development Management

AUERBACH Data Processing Management Library

James Hannan, Editor

•

Contributors To This Volume

Norman Carter
Development Systems International, Studio City CA

———

Jerome E. Dyba
Catonsville MD

———

Steven A. Epner
President, The User Group Incorporated, St. Louis MO

———

Leslie H. Green
The Fidelity Bank, Philadelphia PA

———

Edward J. Kirby
Systems Design Consultant, Sherborn MA

———

Henry C. Lucas, Jr.
Professor and Chairman, Computer Applications and Information Systems
Graduate School of Business Administration, New York University
New York NY

———

G.R. Eugenia Schneider
Naval Weapons Center, China Lake CA

———

James A. Senn
Associate Professor of Management Information Systems, School of Management,
State University of New York, Binghamton NY

———

John Shackleton
ASI Corporate, Arlington Heights IL

———

Stephen P. Taylor
Sperry Univac, Irvine CA

———

Raymond P. Wenig
President, International Management Services Incorporated
Framingham MA

———

A Practical Guide to

Systems Development Management

Edited by James Hannan

AUERBACH Publishers Incorporated
Pennsauken NJ

VAN NOSTRAND REINHOLD COMPANY
New York Cincinnati Toronto London Melbourne

Copyright © 1982 by AUERBACH Publishers Inc

Library of Congress Catalog Card Number 82-11610

ISBN 0-442-20915-0

Printed in the United States of America

Published in the United States in 1982
by Van Nostrand Reinhold Company Inc
135 West 50th Street
New York NY 10020 USA

16 15 14 13 12 11 10 9 8 7 6 5 4 3 2 1

Library of Congress Cataloging in Publication Data
Main entry under title:

A Practical guide to systems development management.

  (Auerbach data processing management library ; 5)
  1. System design. 2. Electronic data processing—Management.
I. Hannan, James, 1946-   . II. Series.
QA76.9.S88P7   1982   658.4'032   82-11610
ISBN 0-442-20915-0 (pbk.)

Contents

Contents

Preface

In its relatively brief existence, the computer has emerged from the back rooms of most organizations to become an integral part of business life. Increasingly sophisticated data processing systems are being used today to solve increasingly complex business problems. As a result, the typical data processing function has become as intricate and specialized as the business enterprise it serves.

Such specialization places a strenuous burden on computer professionals. Not only must they possess specific technical expertise, they must understand how to apply their special knowledge in support of business objectives and goals. A computer professional's effectiveness and career hinge on how ably he or she manages this challenge.

To assist computer professionals in meeting this challenge, AUERBACH Publishers has developed the *AUERBACH Data Processing Management Library*. The series comprises eight volumes, each addressing the management of a specific DP function:

A Practical Guide to Data Processing Management
A Practical Guide to Programming Management
A Practical Guide to Data Communications Management
A Practical Guide to Data Base Management
A Practical Guide to Systems Development Management
A Practical Guide to Data Center Operations Management
A Practical Guide to EDP Auditing
A Practical Guide to Distributed Processing Management

Each volume contains well-tested, practical solutions to the most common and pressing set of problems facing the manager of that function. Supplying the solutions is a prominent group of DP practitioners—people who make their living in the areas they write about. The concise, focused chapters are designed to help the reader directly apply the solutions they contain to his or her environment.

AUERBACH has been serving the information needs of computer professionals for more than 25 years and knows how to help them increase their effectiveness and enhance their careers. The *AUERBACH Data Processing Management Library* is just one of the company's many offerings in this field.

James Hannan
Assistant Vice President
AUERBACH Publishers

Introduction

Systems development has traditionally been considered the heart of the data processing function and one of the most challenging of all DP activities. Proponents of this viewpoint note that developing cost-effective solutions to an organization's business needs requires a rare mix of business, technical, interpersonal, and managerial skills. To be successful, systems analysts need to know as much about an organization's structure, function, goals, and objectives as they do about the latest developments in hardware and software technology. They must be able to interact effectively with different levels of user and DP management in the course of their analysis and design work. And they must be familiar with the array of available development and project management tools and techniques.

Whether one accepts the contention that systems development is the preeminent data processing activity or not, it is difficult to deny that it has become significantly more complex and challenging in recent years. Several factors have contributed to this trend. Users, an increasing percentage of whom are more knowledgeable about and comfortable with computers, are demanding more sophisticated solutions to a greater number of business problems. The business problems themselves have become more complex in the face of intensified competition, a less predictable economic climate, and the reshaping of established patterns of business behavior. Add to these factors a rapidly changing technological environment, and the challenge to develop useful systems on time and within budget becomes formidable indeed. This volume of the *AUERBACH Data Processing Management Library* is designed to help systems developers meet that challenge.

We have commissioned an outstanding group of DP practitioners to share the benefits of their extensive and varied experience in systems development. Our authors have written on a carefully chosen range of topics and have provided proven, practical advice for managing the systems development function more productively.

In Chapter One, Edward J. Kirby discusses the characteristics and management skills that the successful systems development manager should possess. He also outlines the manager's functions and day-to-day activities and points up common problems that the manager is likely to encounter, together with practical solutions.

A major concern of any systems development manager is the establishment and use of a standardized development methodology. Whether developed in-house or purchased from a vendor, standard methodologies help ensure systems reliability, quality, and predictability as well as user satisfaction. In Chapter Two, John Shackleton examines the major characteristics of systems development methodology packages and briefly describes a number of the more popular vendor packages available.

Another important challenge confronting development managers is cultivating and retaining competent project managers. Structured, comprehensive performance appraisals are effective tools for accomplishing that goal. In ''Performance Appraisal of Project Managers,'' Norman Carter describes performance appraisal techniques that help both project leaders and development managers understand evaluations. Also included are procedures, a checklist, and sample forms that can facilitate performance appraisal.

No matter how competent development personnel are, managers sometimes need additional people on a temporary basis to handle excessive work loads or to provide a specialized skill for a project. For such help managers often turn to consultants. In Chapter Four, Steven A. Epner offers practical advice for selecting and using a consultant and provides a sample consultant contract.

Improving the results of the analysis phase of the systems development life cycle and the overall effectiveness of the project team can pay large dividends in the later stages of a development project. A comprehensive systems analysis checklist can help bring about such improvement and can also help produce consistent results while contributing to the expertise of the team members that perform the analysis. Raymond P. Wenig provides such a checklist in Chapter Five.

Many of the failures of computer-based information systems are attributable to their not having been designed with the end user in mind. In his ''User-Oriented Systems Analysis and Design,'' Henry C. Lucas, Jr., discusses analysis and design techniques that ensure the development of quality systems that meet user needs. Stephen P. Taylor then addresses methods for developing a special kind of information system—a decision support system—that is designed around the user organization's decision-making style in Chapter Seven.

Although user-centered analysis and design can help satisfy user demands for better and more responsive systems, managers are often hardpressed to keep pace with the sheer volume of systems that users request. As a result, software packages have become an attractive alternative to developing systems in-house. In Chapter Eight, Raymond Wenig discusses methods for evaluating the internal structure and operational characteristics of software packages and explains how such evaluations can be used in package selection.

Designing a new system or modifying an existing system or package involves the coordination of people, resources, and a nonrecurring set of relatively complex tasks. Such an undertaking requires effective project management if it is to succeed. In ''Organizing for Project Management,'' Leslie H. Green discusses the essential elements of effective project management and alternative project management structures.

A major task of the project team is reviewing the system design for any errors or omissions. The use of structured walkthroughs is a proven technique to "proof" program design, detect errors, and control structure. James A. Senn discusses the concept of structured walkthroughs and describes how to apply it in Chapter Ten.

In addition to conducting design reviews, it is also advisable to perform post-implementation reviews. Such reviews reveal if the systems development process was properly applied and managed and if the anticipated benefits of the new or revised system were attained. In Chapter Eleven, Jerome E. Dyba provides a methodology and checklist for reviewing systems following implementation.

Although a phase of the life cycle that most managers would rather ignore, program maintenance is a costly, time-consuming process that may account for as much as 80 percent of software costs. In Chapter Twelve, G.R. Eugenia Schneider discusses procedures that can help reduce these costs through comprehensive documentation of all maintenance. She describes the necessary maintenance activities, as well as the procedures for performing them, and provides practical documentation formats.

1 The Systems Development Manager

by Edward J. Kirby

INTRODUCTION

Data processing, particularly systems development, is a high-technology activity. It may indeed be the only activity in an organization that must respond to constant, rapid technological changes that demand changes in work habits and organization. An additional problem is that systems development managers, many of whom are promoted from the technical ranks, tend to be naive and inexperienced in such basic managerial skills as budgeting, human relations, and communications.

A problem arises in management goal orientation. Upper management is accustomed to work being completed on time and at close-to-estimated costs. This, unfortunately, is seldom true with systems development projects. Systems for information processing have historically been late and excessive in cost.

Some of the bad reputation is undeserved. The functional goals of other project types tend to be straightforward and specific; those of information systems, elusive and general. Systems developers contribute to this problem by failing to "value engineer" their products and instead trying to achieve heretofore unattained levels of technical perfection. In extreme cases, they include in a system functions and features that far exceed the users' expectations and, needless to say, their budgets as well. They cheerfully accept changes to specifications, at first, then they tolerate them; finally, in a state of panic, they attempt to reject them. Systems developers rarely insist on the increases in cost and development time necessitated by specification changes.

This chapter, which is written mainly for the new manager, discusses how these problems can be handled or avoided. Other problem areas are also discussed, and the characteristics and functions of the systems development manager are described.

CHARACTERISTICS OF A GOOD SYSTEMS DEVELOPMENT MANAGER

The systems development manager must be oriented toward providing service. Computer systems do not exist for their own sake; they are developed as tools to help people work. The system, like any tool, must be matched to the worker as well as the work. The worker is the best authority on himself and usually is one of the best sources regarding his work. He does not, however, know very much about toolmaking. Therefore, a continuing dialogue with system users is essential to successful systems. The service attitude is one of promptness, responsiveness, solicitousness, and readiness to provide help at any time, without constant intervention.

The development manager must be a salesman, able to market to users and his management his ideas and approaches. After they have been approved, he must sell the ideas to his technical staff. Furthermore, he must be persuasive enough to get the resources he needs.

The development manager must be a planner, presenting detailed short-, intermediate-, and long-range plans in writing for all of his group's activities. Still, he must be flexible enough to adapt and alter these plans to accommodate changing circumstances. He must have the temperament to accept change without frustration and the ability to allay the frustrations of his personnel. Foresightedness in effective contingency planning is one of his most valuable assets.

Resourcefulness is the key to obtaining the resources that are unavailable through conventional channels. It is the attribute that enables people to see the alternative solutions to any problem and pursue each of these alternatives until a satisfactory solution is found. Its companion virtue is, of course, the serenity to accept alternatives felt to be imperfect or that may have been conceived by someone else.

The systems development manager must have skill in dealing with others so that he can influence those both inside and outside his development group. He must have credibility, and his staff must be loyal to him. This requires both fairness and equal treatment in work assignments and in the performance he expects. Standard procedures that provide division of less desirable work and objective performance standards are helpful. Sympathy for the staff's problems can be significant in systems development projects, which often involve late hours, difficult deadlines, and the usual frustrations of technical work. It is safe to assume that the secretive and deceptive manager does not gain the confidence of his staff.

Technical proficiency has not been mentioned because it cannot be proved that this is an asset to the systems development manager. Basic management skills and awareness of available DP tools and techniques go a long way toward ensuring a systems development manager's success.

MANAGEMENT SKILLS

Technical Awareness

Systems development managers must maintain up-to-date knowledge of the latest methods, techniques, and products. This responsibility is seldom neglected because most development managers are both relaxed and stimulated by books, periodicals, and professional society meetings. What is neglected is the responsibility for passing this information to staff technicians as well as to upper management. This is especially important because the lay press reports these developments in terms that are obsolete, oversimplified, forbidding, and unrealistic.

Managerial Ability

Another largely underestimated responsibility is that of managing people. Because systems analysts and programmers are, for the most part, self-directed professionals, who are accustomed to receiving mid- to long-term assignments, it is all too frequently assumed that they need little or no management. The falseness of this assumption is proved, of course, by their rapid migration between employers. Top management of organizations with large staffs of other types of professionals seldom understands that one of the principal problems of planning systems development is that one or more key people in any project lasting more than a year will almost surely resign within that time.

Ability to Delegate

All managers must delegate authority. Inexperienced systems development managers who came up through the technical ranks frequently fail in this area. Without delegation of authority, a manager's time is consumed attending to excessive detail, and workers become resentful because they feel that their manager lacks confidence in them. Well-distributed authority in a development project ensures efficient handling of crises and continuity when key people resign. A hierarchy of technical decision levels must be designed so that the systems development manager is free to discharge his other responsibilities.

In addition, the extent of the supervisor's authority should be clear both to him and to his subordinates. The systems manager will gain tremendous savings of time as the result of clarifying and resolving task conflicts. After work segments have been delegated, the manager must, of course, discipline himself to a policy of noninterference.

Ability to Motivate Systems Personnel

Few experienced systems managers would subscribe to the misconception that DP personnel are universally self-motivated, challenged by the risk and excitement of their careers, and dedicated to reaching ever-higher pinnacles of excellence for its own sake. Nonetheless, the attention given to employee motivation is frequently intermittent and sometimes so casual as to be unrecognizable. In motivating systems people, the first concept that must be understood is their work goal. When all theories of motivation are sorted, the development and support of information systems is clearly their objective.

This is the goal of the manager, the group, and all of its members. The objectives of personal development, skills improvement, and work satisfaction are incidental, although not unimportant. Members of the systems development team must strive toward their primary objective under the systems manager's direction and perform to the standards he has set.

The approach many systems managers adopt toward motivation is revealed by their desire to be fair. To them, being fair consists of rewarding good performers and not rewarding the others. Although this method leads to great self-satisfaction, it is of little use in meeting productivity objectives. These objectives can be met only by motivating all members of the systems team toward more and better work. A good practice for the manager who wishes to improve group performance is to devote his managerial attentions more intensively to average and below-average performers.

There is little benefit in negatively motivating professional/technical people. For some, admittedly, fear of punishment will provide an incentive to perform, but a much stronger incentive will be to seek another, more pleasant working environment. For others, fear is ineffective because it distracts them to the point where they cannot function.

Ability to Manage Personal Time

Some managers have an open-door policy, which, unfortunately, works to everyone's disadvantage. These managers are very difficult to meet with because someone else gets through the open door first. The managers themselves have to carry home a briefcase full of work every night to catch up on what they should have accomplished during the day.

The systems manager should keep office hours on an appointment basis. It is not unreasonable to insist that employees request an appointment before meeting with managers. (In true emergencies, of course, this request may come just before the appointment.) By working according to schedule, the manager can tell his visitor in advance how much time he has available.

The manager's schedule should be based on the best overall distribution of time. Typically, a systems manager may decide to spend 25 percent of his time on communications with users and his own management, 35 percent in a staff communications and work direction, and 40 percent in planning and administration. Although the percentages will vary from time to time, they should be used as a guide for preparing all other schedules. Significant deviations are inadvisable.

In scheduling appointments and activities, the manager should use his calendar as a tool to determine adherence to his overall plan of time distribution. This means that he must record the amount of time spent on activities that were not scheduled in advance. If he does not attempt to compensate for deviations from the schedule, he is probably neglecting some area of responsibility. As much activity as possible should be scheduled in advance on a weekly or biweekly basis. This means that the manager should spend from five to fifteen minutes each day on mini-schedules, blocking out anticipated activities on the calendar according to the exact times when they will begin and end.

FUNCTIONS

First and foremost, the systems development manager is a manager—more specifically, a manager of people, policies, and technical efforts. His functions are within several general areas.

Policies. The systems development manager must enforce, and in many cases define and implement, the following types of policies:
- Governmental (e.g., social security taxes)
- Organizational (e.g., expense account reporting)
- Departmental (e.g., personnel reviews)
- Project management and systems development methodology (e.g., status review formats)
- Task level (e.g., program-naming conventions)

Policies are standing guidelines for the performance of management and technical work. They represent the current views of the organization and the systems manager(s) regarding how various work efforts are to be performed.

Policies save management work by eliminating endless explanation and discussions of how relatively minor tasks should be accomplished. They provide a basis for intelligent task clarification. Policies help avoid confusion and open apparent ambiguities to discussion and resolution.

A manager should have firm ideas about the practices and priorities he wishes to implement. They should be thoroughly reviewed before publication and discussed with others to uncover possible misinterpretations and to help gain their acceptance. Whenever a new policy is created, it must be compared with previous ones to avoid contradiction. All new policies should be categorized and indexed.

Policies have no meaning if they are not enforced. It is essential to discipline those who fail to meet departmental standards. Even more important, the manager himself must closely adhere to them.

The policies must be justifiable to the highest level of company management if their support is to be expected when disciplinary action is taken against policy offenders. In addition, it is most important to foster an environment in which policies and changes can be discussed freely and intelligently; otherwise, workers and users may consider them a hindrance to productivity.

Performance Review. A periodic performance review and appraisal should be conducted for each employee.

The principal deterrent to poor performance should be the employee's awareness of his role in group success or failure. After providing direction, the manager's role is to correct any behavior that detracts from good team performance. When a worker does not adequately contribute to team output, the manager's immediate duty is to explain clearly the behavior required. If this does not correct the problem, the manager must investigate the consequences of removing the worker from the team or organization. If removal from the team is not a viable alternative, or if, in the manager's judgment, removal would be even more detrimental than allowing the worker to remain, there is no option other than further explanation and instruction. If, however, removal is considered a sound approach, the manager may wish to warn the employee that if the fault is not corrected, he will be removed from the project group. Beyond this, however, managers must be extremely careful about making remarks that may be construed as either threats or promises. Credibility is a leader's strongest tool.

Arbitration of the Use of Technical Resources. The development manager is the arbiter for technical resource use. He allocates machine time, data preparation time, and specialist time among his projects and assigns and reassigns priorities. He reviews project plans and estimates and decides what revisions are necessary to avoid overlaps and overcommitment of resources. Most important, he must reconcile requirements and desired resources, with available funds.

Training. When the systems manager engages in design, development, or programming, he is ignoring his responsibility as a manager. If others cannot do the work as well as he can, he must teach them or provide them with a means of education so that they can better perform.

All systems people want to improve their skills and increase their value to the organization. A skills improvement program, formal and structured, should be an integral part of every systems department. It should include training of less-experienced people by senior technicians, including, if applicable, the systems manager; cross-training in different activities; and outside training through packaged products, lecturers, and seminars.

Not all training in the systems department should be technical. Because systems are created based on user requirements, communication is an essential element in the development process. Training in oral and written communication skills should be made available at all levels.

Preparation and/or Review of Cost/Benefit Analyses. This process should be conducted on the basis of known facts and limited to management decisions either already made or within one's own power to make. The present or previous cost of the function to be performed by the system can be calculated. The positive benefits of the proposed system can be stated subjectively.

Overseeing Project Management. Project management techniques should be second nature to the systems development manager. Reporting the amount of time spent on projects by team members falls within this function. Time reporting is often more of a detriment to achievement than a benefit to management. Reporting schemes often are so complicated that they waste enormous amounts of time just in their preparation and submission. Requesting employees to report very small time periods leads them to feel a lack of management confidence in them and forces them to fabricate their reports.

Most time reporting systems have no tolerance for any activities other than those directly related to development projects. Many systems also force review at several supervisory levels. This necessitates earlier and earlier submission of time reports. It is ironic that many of the systems that require reporting in one-hundredth-of-an-hour increments are the same ones that force reporting as early as one and one-half days before the end of the period. Systems development time can probably be reported most accurately in increments of one-half day or more.

Problem Solving. It is a fact that responding to problems consumes a large portion of every systems manager's time. There are some common approaches to problem solving that are more satisfactory than others and permit a systems manager more time to conduct his work without undue pressure. The first of these is dealing with expected or suspected problems quickly, for it is well known that problems intensify with age. One aid to early awareness of problems is long-range planning. Thorough long-range planning points up unaddressed areas in which problems are likely to occur or where, historically, problems have occurred. When a potential problem is recognized, the manager should quickly delegate all or part of its solution. Often, time-consuming research or rework is required to solve problems. If these activities are not delegated, the manager will soon face a problem backlog that permits no time for any management activity other than addressing problems.

It is difficult to distinguish between potential problems and nonproblems. Nonproblems are a series of symptoms or indicators that point to a discrepancy or condition that simply does not exist. For example, a user becomes agitated because a description of a required function in his system cannot be found in the user manual. The function exists, but he has either missed the section that describes it or the description is inadequate. These issues are as serious to those who report them as are real problems and are equally deserving of respect and concern. They underscore, however, the need for research and analysis to prevent a problem from becoming a crisis.

Any time an unsatisfactory condition is reported, the correct systems approach is to view it as the affected person views it. Primary emphasis should be on how long it will take for a response or solution to be provided. The affected person has little interest in who will solve the problem or how.

When approaching problems from a management standpoint, the emphasis should be on who will solve the problem. Responsibility should be assigned to an individual, and he should be asked to report on how the problem will be solved. Creating a solution consists of devising a method, testing to determine its feasibility, and then implementing the correction. Systems people should be taught to solve problems calmly and quickly, responding to reported needs rather than reacting to emotionally charged or chaotic situations. No one will insist that the manager personally solve their problem if he can persuade them that he has delegated it to the best-qualified individual.

MODUS OPERANDI

The day-to-day actions of a systems development manager are particularly important because inevitably there are crises that he must face with equanimity. Following a routine can have a calming effect on development activities. There are certain elements that should be included in the routine.

Supervision. Supervision should be casual but frequent. It is inappropriate to scrutinize the activities of professionals at a detailed level, but a close management "presence" can be reassuring and reinforcing.

Supervision on a regular basis can be as simple as a brief visit and a few words of encouragement. The first level of performance evaluation should occur at the time of assignment and completion of tasks. Employees should set their own objectives, with management guidance. An employee in whom the manager lacks confidence should be given shorter-term objectives. The nature of systems and programming work is such that often the technician himself has difficulty measuring his progress, and constantly asking him about it can be confusing as well as aggravating.

Conducting and Attending Meetings. Many systems managers complain bitterly that all of their time is consumed by nonproductive activities. They spend their time in endless meetings with upper management and users. Their administrative tasks are such a burden that no time is left for the technical aspects of their job. They are firmly convinced that no relief is possible because they have no choice but to answer their boss' demands. Furthermore, they see their jobs as problem driven; just when they are about to get organized, another new crisis develops.

The development manager should schedule brief, regular meetings with upper management to inform them of the latest achievements. It is important to hold these meetings when things are going smoothly as well as when there are problems. Users should be met with on a regular basis simply to ask if work is progressing to their satisfaction and to inquire whether there is anything further that can be done for them. This reinforces the image of service, which is essential to systems development success.

When attending meetings conducted by others, the development manager should indicate in advance how much time he can spare for them, and he should arrive promptly. When diplomatically possible, he should leave meetings when the allotted time has expired. Good personal time management is an excellent example a manager can set for his subordinates.

Meetings should be carefully controlled, unless the manager and attendees have spare time. If managers were to estimate the number of man-hours and the attendant cost of each meeting before convening it, they would have far fewer and smaller ones. Meetings should be brief, well organized (with an agenda, of course), run firmly by the chairman, and cancelled when they are no longer necessary.

Reporting. Reporting can be a valuable management tool because reports permit a manager to state problems and accomplishments objectively and with the correct perspective. Reports should be developed in a predetermined format but should be flexible enough to state exceptions. They should have sufficient continuity that if compiled, they would read as the history of a development project or projects. It is obvious that the most readable reports will be the most effective. The major points should always be summarized briefly at the beginning because some readers have neither the time nor the need to read all details.

PROBLEM AREAS

Coping with Design Change Requests. A rigid policy regarding requested changes to systems designs is required for the systems development manager to properly control this area. The only persons with whom specifications changes should be discussed are those with budget responsibility. There is no such thing as a "free" system change because any change requires a revision of system specifications. User management should be apprised of this fact. In the case of a deletion, there may be an offsetting saving, but, nonetheless, the most insignificant changes still cost. Firm policies must also be instituted regarding those from whom changes will be accepted, and a system of formal proposals and acceptances should be developed.

Maintenance. Another problem area for systems development managers is related to the undesirable tasks that must be performed in the course of development work. One of these tasks is program maintenance. Maintenance becomes a serious problem for managers when a programmer of long tenure is assigned this responsibility. This senior person, of course, may be the only person qualified to maintain his programs. It is conceivable that eventually he could reach the point where he would no longer have time to do any development work at all. This indicates a need for more maintainable programs and better documentation—and, especially, a full-time maintenance function.

The need for such a function is obvious if talented individuals are to be kept at high levels of productivity. The question is who will perform maintenance. Maintenance can be used effectively as a training vehicle, but there is the attendant risk of inexperienced hands working on the programs. Another solution is to treat maintenance as a rotating assignment, delegating tasks when they arise to alternate programmers. The difficulty is that maintenance must also be scheduled according to programmer availability; this does not always result in a fair distribution of these less desirable assignments.

Documentation. Closely related to maintenance is program documentation. The better the documentation is, the more easily a program can be maintained. Primary program documentation is the commentary that the programmer codes in the allocated spaces of his source statements. From these a program narrative and flow chart can be prepared; these become the basis for any further documents.

Two flaws that lead to poor documentation exist in this approach. First, the comments written with the source code are detailed and informative when the code is rough and new, but the corrections made in a frantic rush for completion bear either sketchy, minimal comments or none at all. Second, many programmers are not good writers, and few enjoy writing. They view documentation as a necessary evil to be finished as quickly as possible. Usually the documents are not reviewed until someone needs them, and by then, the author/programmer may not be available for explanation. The preceding presents a strong case for the use of a technical writer in any development group large enough to keep one occupied full time. The difference in the quality of documents produced by one who enjoys writing and one who despises it can be justification for such a position.

CONCLUSION

There is an emotional barrier to systems development that stems from resistance to automation. This difficulty has always faced computer professionals, many of whom have become so callous to it they forget about it. Nonetheless it exists and proliferates when tales are told of how computer errors have caused one disaster or another. There is also the fear, of course, of computers replacing people. This is another area where some systems developers must share the blame for their attempts at crude cost justifications, based on jobs they predict can be eliminated.

These factors have contributed to the isolation of systems development departments and their assignment as an outcast position that is detrimental because the results of their work can only be successful if adopted by the mainstream organization. Defensive development managers often not only accept this isolation willingly but encourage it. They feel that if their group is left alone they can accomplish more.

The systems development department, under the leadership of its manager, must join and remain in the mainstream of the organization. Without availing themselves of the opportunity to become as familiar as possible with the character and special needs of the user, the developers cannot be responsive. Without allowing the other members of the organization to observe closely systems under development, the developers are missing a wonderful opportunity to promote system acceptance after installation.

Bibliography

Drucker, Peter F. *Effective Executive*. New York: Harper & Row, 1967.

Mackenzie, R. Alec. *The Time Trap*. New York: American Management Association, 1972.

Stoner, James A. F. *Management*. Englewood Cliffs NJ: Prentice-Hall, 1978.

2 Systems Development Methodology Packages

by John Shackleton

INTRODUCTION

The methodological alternatives available when developing a software system are to purchase a systems development methodology package or to develop one's own methodology in-house. Often the home-grown methodology is successful, as in certain large corporations.

The other alternative, using a vendor-supplied systems development methodology package, requires asking certain questions regarding each package. One question to keep in mind when evaluating a systems development methodology package is whether it provides standardization in the development process that allows management to accurately predict time and resource requirements. One should also determine whether the package provides greater user satisfaction and helps produce a better-quality product. In addition, it is pertinent to consider whether the methodology can be understood and used effectively by inexperienced personnel.

This chapter discusses the major considerations in selecting a systems development methodology package and briefly describes certain packages now on the market.

CONSIDERATIONS IN CHOOSING A SYSTEMS DEVELOPMENT METHODOLOGY PACKAGE

One problem in choosing a systems development methodology package is finding informative literature on the large number of packages available. Many packages are listed in the *Survey of CPM Scheduling Software Packages and Related Project Control Programs* [1]. This reference manual briefly discusses each package and provides vendor addresses. Three areas should be considered when evaluating these packages:

- Organization
- Implementation considerations
- Total cost

Organization of the Package

The methodology should be structured with clearly defined life cycle phases and tasks, with end-of-phase documentation generated as a by-product of each phase activity. The package should give clear examples of all major deliverables and should state exactly the activities of each task and the level of detail required.

Some methodologies break down tasks into minute detail in the hope that inexperienced developers, by completing all the tasks, will produce a better system. A highly detailed methodology requires a large amount of unnecessary paperwork, however, which usually results in a less usable methodology. An average task should take from 10 to 50 man-hours to complete.

The package should provide automated tools or manual guidelines for estimating development costs and time. There are a number of estimating methods to choose from; one or more may be used in a particular package. The formula for estimating can be based on the difficulty of each program, the experience of the personnel available, and so on. The method may estimate from the parts of the system to the whole, or it may use historical information about similar projects. Since history has shown that most systems estimates are too low, any technique capable of enlarging estimates should be encouraged. Estimates done at the detailed task level usually accomplish this.

To establish a basis for measuring the project's progress, the package should provide automated tools or manual guidelines for assigning and scheduling resources. Scheduling can prove a major downfall for most project managers. As systems become increasingly larger in scope and complexity, an automated schedule becomes a necessity.

The package should be adaptable from small to large projects (or vice versa). The analyst should be able to skip some steps on small projects. The package should also be able to handle complex projects and should deal with data base and data communications as well as batch projects.

The package should improve the quality of the system. There are a number of questions to be asked regarding system quality:

- How much will the methodology affect the system in terms of increasing revenue, avoiding cost, or improving service?
- How easy, quick, and inexpensive is it to change the system?
- Will any future changes have a major impact on the existing system?
- Will the methodology provide reports or queries quickly and inexpensively?

Most of the methodologies now on the market do provide many useful management reports that are easy and inexpensive to modify.

Automated Tools. Many methodologies (e.g., PRIDE-ASDM) have a number of automated tools built into them. Others, like STRADIS, use other vendor software packages (e.g., IBM's DATAMANAGER). Virtually all vendors have plans for some sort of automated tool. Some of the automated tools that are or will be available are:

- Project planning/estimating package
- Project control/reporting package
- Data dictionary
- Data base design aid

- Systems design aid
- Graphics support software for documentation
- Test processing for documentation

Implementation Considerations

It appears that most packages take longer to successfully implement than vendors state. Vendor estimates of implementation time range from three weeks to six months. User experience, however, indicates that the implementation often takes from two to three years. Before implementing a package, it is crucial that management have a schedule for making the change. Since most packages require extensive user tailoring, the extent of tailoring should be agreed upon by top management before implementation begins.

Probably the most important factor in successful implementation is proper training. Vendor training varies from one day to six months, with varying degrees of success. Training should include management, users, and the technical staff. The training should cover all aspects of the project cycle and utilize case studies.

Most successful implementations begin with a small- to medium-sized pilot project, carried out by the best and most experienced staff members available. The results of each phase in the pilot project should be carefully documented and the final results presented to top management. After necessary modifications have been made to the package, it should be used on all future systems development projects.

Total Cost

In addition to the cost of the vendor package itself, a number of incidental costs are usually incurred when purchasing a systems development package (e.g., certain customizing, training, and consulting costs). Initial use of a package also usually incurs a cost increase because of the learning curve. Because some packages require extensive documentation, cost increases may be permanent.

SYSTEMS DEVELOPMENT METHODOLOGY PACKAGES

The following are descriptions of the more popular vendor-supplied methodology packages. It should be noted that the source for the number of users of each package is the vendor.

CARA Systems Development Standards

The fundamental philosophy behind the development of CARA Systems Development Standards is to keep the standards as simple as possible. The methodology, developed at Kraft Incorporated in 1977, now has approximately 100 users.

The systems development standards consist of three publications: the reference card, the handbook, and the reference manual. These are organized to facilitate cross-referencing.

The reference card provides an overview of the systems development life cycle by identifying phases, costs, activities, and review and decision points. It also serves as an index for the handbook and the reference manual. The card is very helpful to experienced users of CARA as a checklist to ensure that all aspects of the development cycle have been covered.

The handbook describes in detail the activities to be done, the documentation that should be produced, and the deliverables to be expected at the completion of each phase. The handbook also identifies the person(s) responsible for each task within the various phases.

The reference manual explains how to organize phases and perform the various tasks within the systems development life cycle.

The CARA systems development life cycle has five phases:
- Feasibility study
- Systems design
- Programming and procedures
- Systems acceptance
- Implementation and support

Each phase is further divided into tasks and subtasks that define the participants in each activity and the documentation that should be produced with each task. Nonetheless, the methodology does not drown the technical user in unnecessary paperwork; 13 documentation forms are considered essential for a project.

Profitable Information by Design (PRIDE) Automated Systems Design Methodology (ASDM)

PRIDE-ASDM, developed and marketed by M. Bryce and Associates, of Cincinnati, is one of the older and more integrated packages available. It encompasses project management, data management, structured analysis and design methods, and documentation. There are currently about 1,000 PRIDE users, 30 percent of whom use the fully integrated PRIDE-ASDM package.

The PRIDE-ASDM development cycle is divided into nine phases:
- System study and evaluation
- System design
- Subsystem design
- Computer procedure design
- Program design
- Computer procedure test
- System test
- System operation
- System audit

Each of the phases produces specific documentation that acts as defined benchmarks throughout the methodology. A manual included with the software package provides examples of all major deliverables.

The ASDM portion of the integrated package consists of an Information Resource Manager (IRM) and an Automated Design Facility (ADF). The IRM is the nucleus of the software package and contains the system's data and organizational components—just like a data dictionary. Unlike a traditional data dictionary, however, IRM presents data in a business systems orientation rather than in a DP programming orientation. The IRM can keep track of data throughout a system, no matter how or where it is stored. IRM also provides important management reports for evaluating project status and performance. ADF acts as a computer-aided design tool that the analyst can use during the analysis and design phases.

PRIDE-ASDM also automatically generates systems documentation as a by-product of the analysis and design efforts. The documentation includes design manuals, user manuals, computer run books, and various project activity reports.

Systems Development Methodology (SDM/70) Project Planning and Control System (PC/70)

SDM/70. Developed by Atlantic Software of Philadelphia, SDM/70 is also one of the older systems development packages. It now has approximately 300 users.

SDM/70 consists of nine manuals:
- Summary guidelines
- System requirements definition
- System design alternatives
- System external specifications
- System internal specifications
- Program development testing
- Conversion/Implementation
- Other supporting guidelines
- Estimating guidelines

The manuals provide a step-by-step detailed description of all tasks to be completed within a phase.

In addition to the nine manuals dealing with the systems development life cycle, a number of management manuals provide management with an understanding of the system and also offer procedures for managing the installation of SDM/70.

The SDM/70 development life cycle is divided into nine phases:
- Service request
- System requirements definition
- System design objectives
- System external specifications
- System internal specifications

- Programming documentation
- System testing and integration
- User/Operations guides
- Post-implementation review

Each phase has specific documentation produced as tasks are completed within a phase. Each task has one or more forms that must be completed to provide proof of completion.

PC/70. Atlantic Software developed this automated planning control system for use in conjunction with SDM/70 or as a standalone software package. PC/70 currently has approximately 560 users. It provides a number of report options to assist managers in planning and scheduling (e.g., manpower availability reports, CPM project scheduling bar charts, and resource planning reports). It also generates reports for controlling performance, project monitoring, time and cost accounting, and measurement and evaluation. The reports are aimed at a number of audiences, namely top management, information systems managers, and users and technical personnel.

SPECTRUM-1

SPECTRUM-1, developed by Toellner and Associates, of Los Angeles, is another older package. There are approximately 200 users. The system development life cycle is divided into three phases that are further divided into 13 subphases, as follows:

- Phase 1—Systems definition
  —Master systems plan
  —User requirements
  —Systems definition
  —Advisability study
- Phase 2—Systems design
  —Preliminary design
  —Systems/Subsystems design
  —Program design
  —Programming/Testing
- Phase 3—Systems implementation
  —Implementation planning
  —System test
  —Operations turnover
  —Start-up/Training
  —Acceptance/Wrap-up

The materials provided in the SPECTRUM-1 package are substantial (30 manuals). They consist of long-range planning procedures, systems development guidelines, project planning and control guidelines, documentation standards, and change control guidelines.

With SPECTRUM-1, much emphasis is placed on the implementation of the package. Toellner and Associates firmly believes that vendor packages require substantial tailoring to individual requirements to obtain maximum benefit from the methodology. As part of the SPECTRUM-1 implementation, from one to six months are allotted to tailor all manuals to individual needs. There is also extensive training for executive management, user management, and technical users, as well as training in estimating, scheduling, and quality review. Toellner and Associates strongly recommends introducing SPECTRUM-1 through a pilot project, after which all new projects would use the methodology.

Structured Analysis, Design and Implementation of Information Systems (STRADIS)

STRADIS was an outgrowth of Gane and Sarson's (Improved Systems Technologies Incorporated) Structured Systems Analysis. Analysts should be familiar with Gane and Sarson's Structured Systems Analysis techniques in order to use STRADIS effectively. One of the most recent methodologies to appear on the market, it has approximately 25 users. Like CARA, STRADIS was designed to keep systems development simple and to hold documentation to a minimum.

STRADIS has seven major deliverables:
- Initial study report
- Detailed study report
- Draft requirements statement
- Total requirements statement
- Outline physical design
- Design statement
- Tested code and procedures manual

The STRADIS package consists of a standards and procedures manual, seminar workshops, a reference card, and a number of wall charts and memory aids. It also includes a reference library of 16 books that address a number of topics (e.g., data base design, structured analysis).

The systems development life cycle is represented in STRADIS by a data flow diagram in which analysts and users can clearly identify the project activities. The data flows from process to process, identifying the documentation for each phase.

EVALUATION CHART

The evaluation chart shown in Table 2-1 can be used as a quick reference to evaluate the packages discussed in this chapter. The evaluation scores were based on a survey of 100 systems development methodology package users that was conducted by Advanced Systems Incorporated. The conclusions drawn from the survey are intended as a guide to aid prospective users in selecting vendor packages and not as an absolute measure of the quality of any particular vendor product. The items listed on the chart are those factors that should be considered in selecting any vendor systems development package. The questionnaire has been included in the Appendix.

Table 2-1. Systems Development Methodology Package Evaluation Chart

| Methodology | Systems Development Charactertistics | | | | | Use Characteristics | | | | | | | Bene-fits | | | Cost of Package* $ |
	Detailed Phases	Detailed Tasks	Scheduling Guidelines	Estimating Guidelines	Quality Control	Understandability	Manageability	Transferability	Automated Tools	End-User Impact	Flexibility of Use	Flexibility of Range	Extent of Use	Life Cycle	Saving	
CARA	2	3	2	2	2	5	4	4	2	3	4	3	3	3	3	28,000
PRIDE-ASDM	4	4	4	4	3	3	4	3	4	4	3	4	4	4	2	74,000
SDM/70-PC/70	4	4	4	3	3	2	3	2	3	3	2	4	4	3	2	70,000
SPECTRUM-1	5	4	3	2	3	2	3	3	2	3	4	4	4	3	3	50,000
STRADIS	3	3	3	2	3	4	4	4	2	4	4	3	2	3	3	30,000

Legend:
5 = High   *Approximate average cost
1 = Low

The chart divides each vendor package into four major components:
- Systems development characteristics
- Use characteristics
- Benefits
- Cost

Systems Development Characteristics. The vendor packages are evaluated from their technical aspect, which is divided into five sections:
- Phased deliverables—Is the systems development life cycle clearly divided into predefined phases, with major documentation deliverables for each phase?
- Checklist of tasks—Are all tasks within a phase clearly identified and defined?
- Scheduling guidelines—Does the methodology or tool assist in managing time and resources? Does it identify project progress or slippage?
- Estimating guidelines—Does the methodology or tool provide a step-by-step description of the estimating process for all phases of the systems development life cycle?
- Quality control—Are there effective quality assurance reviews with guidelines built into the methodology or tool for use at appropriate times within the systems development life cycle?

Use Characteristics. The vendor packages are evaluated based on eight aspects of use:

- Understandability—How easily can someone unfamiliar with the methodology or tool understand its results?
- Manageability—How easy is the methodology or tool to manage and control?
- Transferability—How easily can the methodology or tool be taught to someone unfamiliar with it?
- Automated tools—Does the package have automated tools (e.g., graphics software support or text documentation) that can easily be obtained and applied to aid in the use of the methodology or tool?
- End-User impact—Is the output from the methodology or tool easily understandable by nontechnical end users? To what extent do users interface with the development cycle?
- Flexibility of use—How easy is it to tailor the package to existing or future internal standards?
- Flexibility of range—To what degree is the package applicable to simple to complex applications?
- Extent of use—How widely is the package currently used?

Benefits. Two types of benefits are evaluated:
- Life cycle benefits—Does the methodology or tool reduce development time and improve the quality of the system?
- Savings—Does the methodology or tool reduce the cost of system development?

Cost of Package. This figure represents the total package cost, including installation but excluding consulting fees.

CONCLUSION

To obtain maximum benefits from a package, the following steps should be followed:
- After obtaining top management's commitment, a strategy for changing the package should be designed.
- The methodology should be tailored to the organization's standards and requirements.
- Management, users, and technical personnel who are involved in the initial system should be trained.
- The new methodology should be introduced on a small or medium pilot project, using the best available people.
- Both success and problems should be monitored and documented as development progresses.
- The preceding five steps should be reiterated until the package is fully implemented.

Vendor packages that bring standardization to the complex problems of systems development can be very helpful.

Reference

1. Project Management Institute. *Survey of CPM Scheduling Software Packages and Related Project Control Programs,* 2d ed. Drexel Hill PA: Project Management Institute, 1980.

APPENDIX

ASI Questionnaire: Evaluating Systems Development Tools and Methodologies

1. What methodology (e.g., SDM/70, SPECTRUM-1) do you use?

2. What tools (e.g., PC/70) do you use?

Please complete one set of questions for each tool or methodology used in your environment.

How detailed is the methodology or tool (i.e., how many man-hours are required, on the average, for the smallest task)?

Less than 1 hour	( )
1–10 hours	( )
10–50 hours (one man-week)	( )
50–100 hours	( )
More than 200 hours (one man-month)	( )

Other comments:

How flexible is the methodology or tool (i.e., the span of simple-to-complex application to which the methodology or tool can be usefully applied)?

Can be used on (check all applicable boxes):

Very simple applications	( )
Simple applications	( )
Average applications	( )
Complex applications	( )
Very complex applications	( )

Other comments:

Does the methodology or tool enhance the quality of the project or system in terms of:

1. Ease of use—Does the system meet user needs, or will it be sabotaged?
2. Reliability—How often/for how long does the system go down (i.e., how many/how serious are the software bugs when it is turned over to production)?
3. Changeability—How easy, quick, and inexpensive is it to change the software or the data base yet still retain a working system?

4. Performance—Does the system provide reports/queries with acceptable response/speed at acceptable cost?

All four components are built into activities and deliverables.   ( )
Methodology or tool makes some contribution to quality.   ( )
Methodology or tool is not concerned with system quality.   ( )

Other comments:

Does the methodology or tool provide realistic examples of deliverables (e.g., design documentation)?

Every deliverable defined is supported with an example.   ( )
Major deliverables have examples.   ( )
Deliverables are defined but have no examples.   ( )
Deliverables are not defined.   ( )

Other comments:

Does the methodology make doing projects easier or harder (i.e., do the benefits of standardization and control cost anything)?

No cost   Project is much easier with methodology than   ( )
                 without.
               Project is slightly easier with methodology   ( )
                 than without.
Not sure   Maybe some cost.   ( )
Some cost   Project is slightly harder with methodology   ( )
                 than without.
Severe cost   Much more work required to do projects with   ( )
                 the methodology.

Other comments:

Which approach is taken by the methodology or tool?

Formula for cost per program.   ( )
Estimate parts, combine to get whole.   ( )
Historical information about similar projects.   ( )
Similar projects.   ( )
Methodology says nothing about estimation.   ( )

In your opinion, how realistic is the estimation approach taken by the methodology or tool?

Very realistic and usable   ( )
Okay if used by an experienced project manager   ( )
Fine in theory but not useful in our shop   ( )

Other comments:

To what extent does the methodology encourage the use of the following structured methodologies: structured analysis, structured design, structured programming, structured walkthroughs, and top-down development?

Use of all five is mandatory.   ( )
Encourages use but not mandatory.   ( )
Supports some methodologies.   ( )
Says nothing about structured techniques.   ( )

To what extent is project/system documentation produced as a by-product or integral part of necessary project work or produced in addition to project work or after the fact?

Completely a by-product   ( )
Mainly a by-product   ( )
Some by-product, some after the fact   ( )
Mainly after the fact   ( )
Completely after the fact   ( )

Other comments:

What training is provided by vendor?

One to four weeks in-house   ( )
Four to eight weeks in-house   ( )
More than nine weeks in-house   ( )
Other: ( )

How easy to learn is the methodology or tool?

Very easy   ( )
Easy   ( )
Average   ( )
Difficult   ( )
Very difficult   ( )

How adequate is the training provided?

Adequate   ( )
Acceptable   ( )

Poor   ( )
Nonexistent   ( )

Other comments:

How long did the methodology take to implement?

Less than six months   ( )
Less than one year   ( )
Less than two years   ( )
More than two years   ( )

Other comments:

What automated tools are available to support the methodology?

| | From Vendor | | From |
	Announced	Delivered	Other Vendor
Project planning/estimating package	( )	( )	( )
Project control/event-time reporting package	( )	( )	( )
Data dictionary	( )	( )	( )
Graphics support software for documentation	( )	( )	( )
Data design aid	( )	( )	( )
Text processing for documentation	( )	( )	( )

Other comments:

Estimate the total cost of installing the methodology.

Cost of documentation packages   _____
Customization by vendor   _____
Customization by your staff   _____
Training—vendor costs   _____
Training by your staff (time, travel)   _____
Consulting—vendor   _____
Other—costs   _____

Other comments:

What do you feel is the most desirable aspect of the tool or methodology?

What do you feel is the least desirable aspect of the tool or methodology?

③ Performance Appraisal of Project Managers by Norman Carter

INTRODUCTION

Appraisal of a project manager's contribution to the development of his staff and to the quality of production is often overlooked by management. Some supervisors apparently feel that a raise and an occasional pat on the back obviate the need for formal evaluation. This attitude, however, can greatly contribute to employee discontent and high turnover.

Companies that conduct regular feedback interviews six months after employees leave have found that lack of effective performance appraisal ranks high on the list of reasons for leaving. In many cases, it is more important than the financial motivation so often discussed at the time of leaving. If lack of effective performance appraisal is indeed a major reason for employee turnover, there are straightforward ways to attack the problem.

There is another reason for conducting regular, formal performance appraisal of project managers: both the Equal Employment Opportunity Commission (EEOC) and Affirmative Action (AA) require that a company be able to demonstrate a direct and traceable relationship between a job description, performance criteria for the job, appraisal of performance against the description and criteria, and direct involvement of the individual in setting, monitoring, and measuring objectives.

Objectives of Performance Appraisal

The primary objectives of performance appraisal are to:
- Review employee progress in terms directly related to the organization and to the individual's job family and position.
- Review and establish measurable performance goals for the next given time period.
- Design objectives, action plans, and training curricula for each individual for current and future job responsibilities.

Note that justification for a requested salary increase is not among these objectives. In fact, a combined performance and compensation appraisal detracts from the objectivity of the performance evaluation (unless, as described under Weighted Performance Goals, the two are inextricably bound); the functional manager may find that he must make unsupported statements or statements that do not reflect a consistent view of the individual's contribution to the department in order to support a requested increase.

Performance appraisal provides the framework within which the growth of an employee can be evaluated independently of the availability of money to compensate that individual. In fact, consistent appraisals are one lever a functional manager can use to correct salary grades or ranges with the compensation manager. Once-a-year fudged performance appraisals make correction of salary inequities almost impossible.

Performance appraisal can also be used for mutual discussion of the professional and technical achievements of the project manager. Performance objectives can be negotiated, thus avoiding unilateral goal setting by the manager.

As part of management's responsibility to project manager development, there must be a willingness to set objectives that permit the project manager maximum freedom to accomplish the job through project team members. This might entail creating different work schedules, changing processes or procedures (with good reason), and/or establishing specific objectives and rewards to encourage improved performance or productivity. In as many ways as possible, the manager must view and appraise the project managers and teams as proprietors of their businesses. This requires the manager to generate a high degree of confidence to the project manager as well as constant and consistent coaching.

Managers as Coaches Rather Than Umpires

The role of a manager can be likened to that of the coach of a team. Each player (or project manager) is taught what to do and how to do it under normal circumstances. As the game proceeds, the coach makes minor adjustments. A coach who does not modify the game plan in response to the play is usually neither respected by his players nor successful in developing or maintaining a winning team.

At the same time, a player is responsible for calling a time-out to discuss a situation observed on the field so that the coach can offer further assistance. In this sense, the success of the team is as important as the success of each individual.

Performance appraisal involves the functional manager in the coaching or counseling of project managers in their overall development rather than just as an umpire dealing with disputes and disruptions. It also involves evaluation of a project manager's coaching and counseling of project team members.

TYPES OF PERFORMANCE APPRAISAL

Performance appraisal usually is viewed as a single activity: sit down, fill out the form, conduct a cursory everything-is-all-right discussion, sign the form, and get back to work. It is not as simple as that, however.

Employees traditionally fall into three categories:
- High performers with high potential
- Average performers
- Marginal performers

All three types of performance can be observed in project managers.

When evaluating a project manager, it is necessary to differentiate between technical capability, skill, and performance and managerial skill and performance. Specific attention to marginal aspects of the job must be evaluated. For example:
- Is the project manager effectively interpreting user and senior management requirements so that objectives for himself and his staff can be properly set?
- Is he capable of recognizing when a performance requirement cannot be attained?
- Is he spending enough time training and counseling project members?

If the functional manager observes that managing skill is deficient, he must ask:
- Is the cause lack of knowledge that can be gained through training?
- Is the cause lack of direction from management?
- Is poor time utilization by the individual creating an imbalance between technical and supervisory performance?

The evaluator should prepare differently for appraisals in each performance category to provide maximum benefit to the project manager and to the company.

The high performer should, of course, be expected to accomplish more than the low performer and also to perform more job-related self-development activities outside of work. A low performer, however, who is to be separated from the company, may be on the job longer than expected while a replacement is obtained. It may be best to remove this type of project manager from a position of responsibility. No training activities should be scheduled for this individual. See Figure 3-1 for a breakdown of performance/training activity requirements.

Training Activity \ Performance Category	High Performer		Average Performer		Marginal Performer	
	Ready	Future	Short Plan	Long Plan	Keep	Separate
Self-development	High	High	Above average	Average	Low	None expected
Classes/ workshops/ seminars	To round out knowledge	Key subjects for next position	Next required missing skills	Selective skill for advancement	To maintain skill	—
Coaching/counseling	Intensive	Intensive	To supplement skills	To prepare for advancement	To maintain skill	To maintain minimum skill until separation
Involvement in other company activities	High	High	Some	As Available	Minimum	—

Figure 3-1. Performance/Training Activity Requirements

PREPARING FOR THE PERFORMANCE APPRAISAL

An effective performance appraisal is a most demanding and rewarding activity, but it requires time—which managers often claim they lack for such appraisals. Time can always be found, however, to interview new hires, to correct work if employee objectives have been poorly set, or to provide training when lack of knowledge causes errors. Often, more time is required to correct a performance problem than to conduct an appraisal, set objectives, and help the employee understand them. Preparing and conducting a thorough, effective performance appraisal should take less than five hours per person.

Review and Evaluation of Performance. This step involves gathering the tools for appraisal, reviewing objectives and accomplishments, considering why things were or were not done as agreed, reviewing the project manager's overall performance, and identifying the individual's strengths and weaknesses. This crucial preparatory activity should take from one to one and one-half hours per person.

Discussion. After the manager and the project manager have prepared for the evaluation, performance, productivity, and continuing objectives should be discussed. This should also take from one to one and one-half hours per person.

Negotiation. If performance evaluation is done consistently and at logical checkpoints throughout the year, differences of opinion should be minimal. Several discussions may be necessary, however, to reach mutually agreeable performance objectives. These discussions may require two meetings of about an hour each.

Completion. Completing and submitting all paperwork in accordance with company procedures should take about 15 minutes.

The Tools

The types of tools discussed in the following sections facilitate performance appraisal.

Standard Forms and Procedures. If standard forms and procedures have not been specified by the company, they should be developed and used consistently. This requirement becomes increasingly important as EEOC and AA continue to expand their roles as protectors of employee rights. Standardization also helps avoid government audits that occur when individuals feel that varying standards are being applied.

Position Descriptions. The project manager job description should be written in specific terms detailing what is to be done and how, in addition to providing broad statements of responsibility and authority.

Job Standards. Job standards and tools should describe project requirements, system development standards and guidelines, departmental standards and policies, and pertinent company policies and procedures.

Assignments/Results. The objectives for the period should be available for review, as should a list of assignments that may have facilitated or impeded achievement of the objectives.

Previous Appraisals. Several prior appraisals should be available for review to help detect such trends as failure to meet objectives or exceeding objectives frequently.

Setting the Meeting Date

To ensure that both parties are effectively prepared, the project manager should receive copies of the performance evaluation forms and instructions at least one week before the discussion date. If special or additional goals have been included, they should be reviewed and communicated to the project manager at this time (preferably in writing). Self-assessment aids can also be made available at this time for the individual to use, if desired.

THE PERFORMANCE APPRAISAL DISCUSSION

At best, performance appraisal begins as a stressful interview. The participants bring different expectations. Until it is understood that their differences are professional and not personal, that compromise need not be all on one side, and that effective negotiation is a sign of professional maturity, the discussion will achieve less than optimal results. The following suggestions should help alleviate the threatening aspects of the discussion.

The Environment. Do not conduct the discussion in a noisy environment or with other people present. For example, do not hold it in a restaurant where customers and serving make communication difficult. (In addition, it is difficult to enjoy a meal under the constraints of such a critical activity as performance appraisal.)

The best setting is a neutral environment (e.g., a conference room) where both parties can come from behind their desks. In addition, try to ensure that the discussion is not interrupted; telephone calls should not be taken by either person during the discussion. Behaviorists state that each time a discussion is interrupted, regaining the concentration and flow that existed before the interruption takes between five and ten minutes.

The atmosphere should be as comfortable as possible. If the atmosphere of the department is shirt sleeve, keep it that way. Do not set up artificially formal barriers. Have some liquid (coffee, soft drink, water) available.

The Discussion. The process must be a discussion, not a monologue. Both parties, but especially the manager, should practice active listening techniques. Notes should be taken and, whenever necessary, read back so that both parties understand and agree on what has been discussed.

Negotiating. When differences of opinion on performance arise, the manager should be prepared to use conflict resolution skills. Resolutions must be within the scope of and consistent with the performance appraisal tools mentioned earlier. Agreements reached outside these constraints, unless carefully documented and well understood, often lead to additional conflict. They are, therefore, self-defeating as a means of improving performance.

Legal Requirements

Although all of the EEOC rulings and AA requirements cannot be detailed in this chapter, the following points should not be overlooked:
- Compliance with the laws is compulsory, not voluntary.
- Intent to follow the laws is not sufficient.
- Documentation of appropriate procedures and policies is required in case of audit.
- The responsible organizations have stated that audits of compliance will be conducted more frequently than in the past.

Not only do these points apply to the performance appraisal of a project manager but, as will become clear in the next section, a project manager must understand and abide by these requirements.

Goal Setting

Two types of objectives setting are required for performance evaluation: qualitative and quantitative.

Qualitative Goals. Too often, all of the established goals are qualitative and include such statements as:
- Will maintain a level of production consistent with the average achieved by other project managers
- Will comply with procedures established by management

Although some qualitative goals can be beneficial, they should be expressed in concrete terms so that the individual understands exactly what is expected. For example, more explicit qualitative goals might be:
- To conduct a workshop with project personnel, within one week of the beginning of a project phase, on the system standards to be applied during that phase. The project manager will report to management (in writing) the date on which the workshop occurred.
- To understand and ensure compliance by all assigned project personnel with company attendance reporting requirements.

Qualitative goals should be kept to the minimum consistent with the assumption that the employee knows the general requirements of the company and his job.

Quantitative Goals. As much as possible, performance goals should be quantitative and restricted to an attainable number, generally between three and five. With more than five goals, activity and accomplishment tend to become too diffuse and judgment imprecise. Spreading fewer than three goals over a similar period of time tends to make recalling sufficient detail difficult.

At a minimum, a quantitative goal should include the following elements:
- A description of the task to be done
- A definition of the standard to be used
- A breakdown of the task into deliverable items and the standard for each; for example:

    To list the eight laws and executive orders that govern EEOC and AA compliance requirements. Within six months the project manager will report to management that the project is in compliance.
- A statement of the value to the individual in meeting the goal; for example:

    Completion of this objective will be valued at 20 percent of the next appraisal. Failure to complete the project within six months may be considered cause for relieving the project manager of his supervisory responsibilities.

    (Note that the reason for the significant penalty in this example is the exposure of the company to legal action if compliance with EEOC and AA regulations is not achieved.)

With project manager objectives, those variables that may cause failure to meet goals must be carefully identified; otherwise, the tendency is to blame something or someone else for the unmet objective. References to signed approvals, accepted specifications, and individuals who must sign off on performance are more necessary at this level than at most others. The project manager should be expected to identify many of these constraints.

JUDGING REWARDS AND PENALTIES

An effective challenge to individuals to improve their performance requires rewards and penalties. Often, the reward is more money and the penalty less, with a range of 3 to 6 percent. In view of today's economy, this may not be sufficient motivation. Rewards not exclusively tied to money should be used.

Weighted Performance Goals. Once agreed-upon objectives are accepted as the normal, expected performance, the effect of other-than-normal performance can be judged. Weighted goals, which define other than standard performance, can be expressed as follows:
- The objective is to complete the project on the schedule described and within a budget of $X, over which you have control. Upon completion, your performance reward will be:

    —On schedule, below budget = normal increase + 10% of budget saved

    —Before schedule, below budget = normal increase + 25% of budget saved

—After schedule or over budget = no increase
- The objective is to implement the XYZ software package successfully and in accordance with the vendor's contract terms and planned schedule and to achieve a level of user satisfaction so that fewer than four complaints will be received by management in the first three months of operation.
  —Should this occur, 50 percent of your performance award will be earned.
  —If the schedule is missed by more than one month or if user complaints exceed four in that period, the performance award will be decreased to 35 percent.
  —If the schedule is missed by more than three months or if complaints exceed 10 in that period, the goal will be considered unmet.

These examples show that while weighted goals expedite quantification of rewards, they require considerable thought, precise definition, and tough-minded enforcement. In most cases, however, a demanding atmosphere, coupled with fair and firm goal setting and evaluation, benefits the individual and the company.

Additional Techniques

Three additional techniques can be used to make performance appraisal more effective. *Totem poling*, *tie breaking*, and *ranking* aid in weighing individuals against each other; these techniques are perhaps most beneficial in situations where resources and opportunities are limited.

Totem Poling. Totem poling is the listing of all employees in order of performance, top to bottom. The totem pole is constructed from the manager's empirical judgment and is then refined by the performance appraisals. Inconsistencies in judgment at appraisal time are minimized since the person constructing the totem pole must ask:

Why have I placed this project manager in this position? Is this placement consistent with the performance appraisal rating?

Tie Breaking. Some form of tie breaking is required when two or more project managers seem to have identical ratings and only one can be selected for advancement. Pertinent rating questions can be developed, with the value of each determined on a basis acceptable to all managers involved in the selection process. Figure 3-2 shows the kinds of questions and value ratings that can be created.

With this tie-breaking technique, each individual is rated and the score is calculated by multiplying the numeric value of the answer by the value rating and then adding all rated items. The result can be used as one input to help break a tie.

Ranking. Totem poles of all project managers in an organization (or department) can be combined for similar job families or project groups. Using a master ranking list, management can:

Individual Rating	Value Rating
1. Demonstrated ability to bring projects in on time and within budget (±5%) 3 Usually better 2 As planned 1 Usually misses	× 3
2. Adherence to SDLC process, stated guidelines, project (job procedure) 3 Always 2 Satisfactory 1 Fails to comply	× 1
3. Effective user relationships (does not require manager intervention) 3 Fewer than two complaints/yr 2 Three to five complaints 1 More than six complaints	× 3
4. Quality production 3 Consistently above standard 2 Meets standard 1 Below standard	× 2
5. Quantity Production 3 Consistently above standard 2 Meets standard 1 Below standard	× 2
6. Meeting agreed-upon objectives 3 Usually betters performance 2 Meets at least 2 out of 3 1 Rarely meets	× 1
7. Making creative input outside of assigned project area 3 Often (2 to 3 times/yr) 2 Sometimes (1/yr) 1 Rarely	× 1
8. Applies training received, when back on job 3 Always 2 Sometimes 1 Rarely	× 1
9. Consistency and accuracy of project planning and estimating 3 Plan always met (barring outside intervention) 2 Plan met 80% of time 1 Plan met less than 50% of the time	× 3
10. Knows and actively supports management objectives 3 Always 2 Usually 1 Rarely	× 2

Figure 3-2. Typical Tie-Breaking Questions

- Identify evaluation inconsistencies among departments or managers
- Identify candidates:
  —For advancement
  —For evaluation of low performance
  —Who are expected to change ranking position during the next 12 to 24 months

PERFORMANCE EVALUATION AND PLANNING PROCEDURES

Each project manager's job performance should be evaluated regularly. This evaluation becomes part of the project manager's personnel records and is a factor in compensation, promotion, training, transfer, and termination. The forms shown in Figures 3-3 through 3-13 can be used in preparing for and conducting performance evaluations.

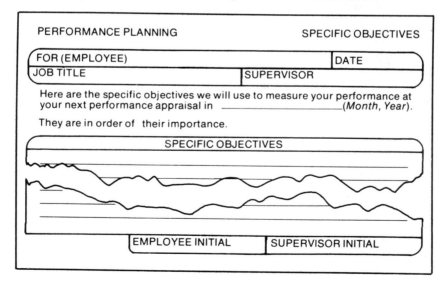

PERFORMANCE PLANNING   JOB STANDARDS

FOR (EMPLOYEE)   DATE

JOB TITLE   SUPERVISOR

Here are the job standards we will use to evaluate your performance at your next performance appraisal in _____(Month, Year).

They are in order of their importance.

STANDARDS

EMPLOYEE INITIAL   SUPERVISOR INITIAL

Figure 3-3. Performance Planning Worksheet: Job Standards

PERFORMANCE PLANNING   SPECIFIC OBJECTIVES

FOR (EMPLOYEE)   DATE

JOB TITLE   SUPERVISOR

Here are the specific objectives we will use to measure your performance at your next performance appraisal in _____(Month, Year).

They are in order of their importance.

SPECIFIC OBJECTIVES

EMPLOYEE INITIAL   SUPERVISOR INITIAL

Figure 3-4. Performance Planning Worksheet: Specific Objectives

PERFORMANCE PLANNING   COMMON PERFORMANCE FACTORS

FOR (EMPLOYEE)	DATE
JOB TITLE	SUPERVISOR

We will consider the common performance factors checked here in monitoring and evaluating your job performance. These will be considered in addition to, not a replacement for, job standards and objectives.

(NOTE: Only check the most important factors. Use the comment section to further explain level of performance expected and the relative importance of each to overall performance on the job.)

	COMMENTS
☐ QUALITY — of finished work regardless of amount completed. Accuracy, neatness, thoroughness.	
☐ QUANTITY — amount of satisfactory work completed. Volume of output, speed in completing assignments.	
☐ TIME MANAGEMENT — meeting deadlines. Utilizing time effectively for maximum output and/or highest quality. Punctuality. Attendance.	
☐ ORGANIZATION — logically plans and organizes own and/or others' work for most effective handling or reduction of unnecessary activities.	
☐ COMMUNICATIONS — effectiveness of written, oral, listening skills.	
☐ KNOWLEDGE OF OWN JOB — know-how and skills necessary to do the job. Adequacy of practical, technical, or professional skills and experience.	
☐ KNOWLEDGE OF RELATED AREAS — awareness of work relationships with other areas.	
☐ LEADERSHIP — ability, skills in orienting, motivating, guiding others. Serving as a good example. Optimum use of staff, other resources to complete task, achieve a goal.	
☐ SELF-DEVELOPMENT — awareness of own strengths, weaknesses, interests. Plans for elimination of deficiencies, attainment of goals. Accepts/seeks new responsibilities.	
☐ SELF-STARTER — working with limited supervision or direction. Following through on own initiative.	
☐ HUMAN RELATIONS — effective work relations with supervisor, peers, others outside working unit, favorable customer relations.	
☐ PLANNING — setting objectives, budgeting, scheduling, forecasting.	
☐ DECISION MAKING — making prompt decisions considering relevant factors and evaluating alternatives.	
☐ COST AWARENESS — awareness of financial impact of decisions, actions. Good business judgment.	
☐ DEVELOPING PEOPLE — recognizing growth potential, development of opportunities, skill in coaching and counseling. Fair and consistent use of discipline. Respect for the individual.	
☐ PERSONNEL PRACTICES — effective and appropriate use of salary and benefits programs, performance appraisal, internal placement, career planning, training and development opportunities, etc.	
☐ AFFIRMATIVE ACTION — working with others harmoniously without regard to race, religion, national origin, sex, age, or handicap. Seeking ways to achieve organizational EEO objectives and timetables. Actively seeking to enhance career objectives of minorities, women, and handicapped people.	
☐ SUPPORT OF SOCIAL POLICY, CONSUMER AFFAIRS PROGRAMS — professional, community, or volunteer activities that promote company objectives. Actively promoting Affirmative Lending and other consumer programs.	
☐ OTHER —	

Figure 3-5. Performance Planning Worksheet: Common Performance Factors

PERFORMANCE PLANNING

These are the revisions, additions, or deletions we have made and the date of change.

_____
_____
_____
_____
_____

Figure 3-6. Performance Planning Worksheet: Negotiated Objectives

A performance evaluation is a communication tool in that project managers are involved in planning their work, targeting performance goals, and measuring results. This allows project managers and their managers to discuss job performance (as it relates to the desired results) openly. It encourages the discussion of career aspirations and the development of plans toward their realization. It enables the development manager to evaluate the project manager's job performance objectively in terms of the position requirements and other negotiated objectives.

Project Manager Performance Categories

Explicitly defined terms, such as the following, should be used in describing an employee's level of performance:

- New in Position—This category includes project managers who need more training and/or experience to achieve basic competence levels. A project manager should remain in this category until performance and productivity increase through experience. A maximum of three months is suggested.
- Marginal—This category includes project managers whose performance needs improvement to achieve basic competence levels (i.e., the performance does not meet minimum job standards or negotiated objectives). The expected results have not been achieved. Improvement to a competent performance level within a reasonable time is required for the project manager to continue in the position.
- Competent—This is the standard level of fully adequate performance (i.e., the project manager's performance meets the previously negotiated objectives). Project managers in this category consistently discharge all job requirements in an able manner, and the expected results are achieved.
- Commendable—This category includes project managers whose job performance exceeds the previously negotiated objectives. The commendable project manager is clearly above average in meeting requirements; better-than-expected results are consistently achieved.

PERFORMANCE PLANNING   INTERIM PERFORMANCE REVIEWS

FOR (EMPLOYEE)   JOB TITLE

FIRST REVIEW   DATE _____

EMPLOYEE INITIAL   SUPERVISOR INITIAL

SECOND REVIEW   DATE _____

EMPLOYEE INITIAL   SUPERVISOR INITIAL

THIRD REVIEW   DATE_____

EMPLOYEE INITIAL   SUPERVISOR INITIAL

Figure 3-7. Performance Planning Worksheet: Interim Reviews

- Distinguished—Project managers in this category have proved themselves to be exceptional in surpassing objectives. Such project managers are outstanding performers whose achievements are readily apparent. They are thus ready for promotion or added responsibilities at an early time.

PERFORMANCE PLANNING

The Performance Planning Interview. The manager should prepare for the interview by reviewing:
- The project manager's position definition.
- Organizational objectives—This review aids in determining which project manager accomplishments are necessary to achieve organizational objectives.
- Appropriate documents prepared by the project manager on the job.

The Performance Planning Worksheet. The worksheet should be completed as follows:
- The development manager and project manager should discuss and then list the job standards, in order of importance, that will be used to evaluate his or her performance (see Figure 3-3).
- Specific objectives that should be met by the project manager should be discussed and listed, also in order of importance (see Figure 3-4).
- Common performance factors (i.e., those not related to specific jobs or departments) that are significant for this project manager should be checked off (see Figure 3-5); appropriate comments should be added.

Quarterly Reviews. When quarterly reviews are necessary or desirable, the development manager should review the Performance Planning Worksheet in order to gauge the project manager's progress toward achieving the stated goals. The project manager should be notified of the review and its expected content at least 24 hours in advance. The following should occur during the review:
- Objectives and desired results should be discussed. If altered circumstances require changing the objectives, new or modified objectives should be inserted at this time (see Figure 3-6).
- The development manager and the project manager should discuss the progress made and complete the appropriate section on the worksheet (see Figure 3-7).

The Performance Planning Worksheet is usually retained within the department after this review.

PERFORMANCE APPRAISAL

The performance planning interview, at which objectives should be negotiated between the project manager and the development manager, should be held within three weeks of the last evaluation (these activities can, of course, be done together). The completed Performance Planning Worksheet should be forwarded within one week to the DP manager, Personnel, and other appropriate departments for review. The worksheet should then be returned to the development manager.

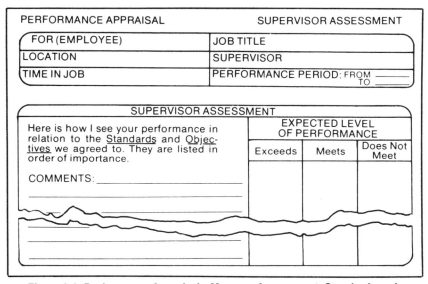

Figure 3-8. Performance Appraisal—Manager Assessment: Standards and Objectives

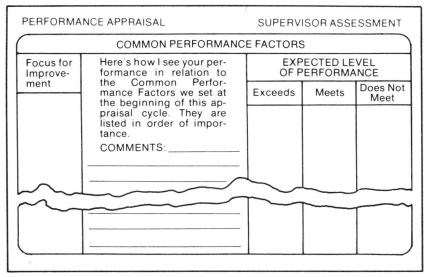

Figure 3-9. Performance Appraisal—Manager Assessment: Common Performance Factors

PERFORMANCE APPRAISAL   SUPERVISOR ASSESSMENT

Here are what I see as your major strengths and abilities, the things you have
done particularly well, and the significant improvements you have made since
your last appraisal:

I think improvement in these areas will increase your overall effectiveness on
the job: (Explain)

I also considered these additional factors (if any) in reaching the overall rating
for you:

OVERALL PERFORMANCE

Here's how I rate your overall performance, based on the performance criteria
we established and considering the relative importance of each:

DOES NOT MEET   MEETS EXPECTED   EXCEEDS
EXPECTED LEVEL   LEVEL OF   EXPECTED LEVEL
OF PERFORMANCE □   PERFORMANCE □   OF PERFORMANCE □

Figure 3-10. Performance Appraisal—Manager Assessment and Rating

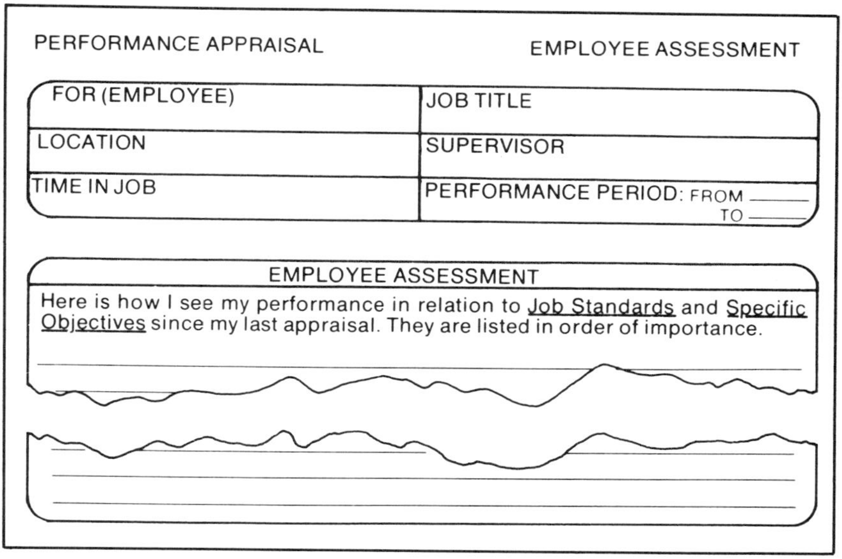
PERFORMANCE APPRAISAL EMPLOYEE ASSESSMENT

FOR (EMPLOYEE)	JOB TITLE
LOCATION	SUPERVISOR
TIME IN JOB	PERFORMANCE PERIOD: FROM ____ TO ____

EMPLOYEE ASSESSMENT

Here is how I see my performance in relation to Job Standards and Specific Objectives since my last appraisal. They are listed in order of importance.

Figure 3-11. Performance Appraisal—Project Manager Assessment: Standards and Objectives

The Appraisal Form

One week before the scheduled evaluation, the project manager should receive a copy of the Performance Planning Worksheet and a copy of the position description; both documents should be brought to the discussion. The development manager should complete the appropriate sections on the Performance Appraisal form prior to the interview. The evaluator should compare the results expected (as indicated on the Performance Planning Worksheet) to the achieved results (see Figures 3-8 and 3-9).

Other factors that the evaluator might consider are absences, outside job-related activities, time management, human relations, and such administrative skills as planning, leadership, organizing, and controlling (see Figure 3-10). The overall performance rating (as shown in Figure 3-10) should be the criterion later used to recommend merit increases. The rating should be based on a comparison of the achieved results with the expected results. The evaluator should emphasize the project manager's strengths and abilities in relation to his or her job performance (see Figure 3-10). He or she should comment on areas in which the project manager can upgrade his or her current performance rating and/or be considered for additional responsibilities.

During the discussion the following should occur:

- The evaluator should consider the project manager's own assessment (see Figures 3-11 and 3-12) in terms of improving his or her effectiveness in the current position as well as possibly developing the project manager for advancement (see Figure 3-13).

- The project manager should write any additional comments concerning the evaluation (see Figure 3-13).
- If there is not sufficient time to prepare a Performance Planning Worksheet for the next period (see Figures 3-3 and 3-6), the evaluator and project manager should schedule a time within the next three weeks in which to do so.

PERFORMANCE APPRAISAL   EMPLOYEE ASSESSMENT

I have shown greatest strength or improvement in performing my job in these areas:

_____

_____

_____

_____

_____

I would like to improve my performance on the job in these areas:

_____

_____

_____

_____

These are my objectives for this job, or for a career, or for my own improvement, for now and in the future.
OR: ☐ At this time, I am satisfied in my current position and wish to remain.
(NOTE: This section is optional. By noting your interests, even if they change later on, your supervisor can provide counseling and direction to help you reach your goals.)

_____

_____

_____

Here are ways that would help me improve my performance or meet my objectives (e.g., more or different help from your supervisor, special training in basic or new skills, cross-training in other areas, etc.).

_____

_____

_____

_____

**Figure 3-12. Performance Appraisal—Project Manager Assessment:
Strengths and Objectives**

PERFORMANCE APPRAISAL   DEVELOPMENT/COMMENTS

DEVELOPMENTAL PLAN

I think we should take these steps to improve your performance on the job, or to help you progress toward your personal career objectives.
(Use career planning tools if appropriate. If the employee wants to remain in the present assignment at this time, please say so here.)

EMPLOYEE COMMENTS

What do you think about this appraisal?

EMPLOYEE SIGNATURE   DATE

(Signature indicates you have seen and discussed this appraisal with your supervisor. It does not necessarily imply agreement with the appraisal or overall rating.)

SUPERVISOR'S SIGNATURE   DATE

REVIEWED BY   DATE

ADDITIONAL REVIEW — (If any)   DATE

Figure 3-13. Developmental Plan and Project Manager Comments

Processing the Performance Appraisal Form

The Performance Appraisal form should be routed to Personnel and other appropriate departments within two days after the interview. The Performance Planning Worksheet covering the period evaluated should be attached.

CONCLUSION

Regular performance appraisals, using the methods discussed in the first part of this chapter and the standardized procedures and forms recommended in the latter part of this chapter, can significantly help project managers understand how well they are performing their jobs and how they are perceived by their managers. As mentioned, the lack of this information is frequently an important factor in employee dissatisfaction and subsequent resignation.

Such evaluations require time and effort to prepare and execute; the benefits to project managers, their managers, and the organization, however, can be substantial.

4 Using a Systems Consultant

by Steven A. Epner

INTRODUCTION

Certain steps are necessary to ensure productive and cost-effective use of systems consultants. These steps involve determining whether the use of a consultant is appropriate and evaluating both the consultant's skills and the organization's needs to ensure that they match. This chapter addresses these issues in six major sections:
- Preliminary considerations
- Establishing and defining deliverables
- Timing and cost constraints
- Locating consultants
- Selecting consultants
- Contracts

PRELIMINARY CONSIDERATIONS

Consultants have varying degrees of skill and experience but share the common goal of providing organizations with temporary assistance for specific needs. A consultant can be defined as "a person who gives expert or professional advice" and " . . . has an assured competence in a particular field or occupation."

This definition raises a major question in data processing. The field does not have a well-defined standard body of knowledge. In addition, many methods may be available to accomplish a given goal. One procedure may be more appropriate than another, but none may be deemed wrong. Competence and expertise thus become difficult to determine.

A consultant's social skills are also important. DP is a field where the interaction between people and machines can make the difference between success and failure. An average system that takes into consideration the man-machine interfaces will often succeed, while the more technically elegant design can fail if it ignores such nontechnical areas. Successful consulting requires both technical and interpersonal skills.

One other critical element is client commitment. No consultant can be expected to work in a vacuum. Successful consulting engagements always include a client liaison who has the responsibility and the authority to act on behalf of the client. Without this liaison, the consultant may be missing the key element necessary in solving the problems or providing the solutions he or she was hired to supply.

There are three major reasons for using a consultant:
- Peak load
- Special skills
- Objectivity

It is important to understand what is involved in each of these situations. Consultants appropriate in one environment may not be useful in another.

Peak Load. Sometimes an organization finds itself committed to completing more work than is possible with in-house resources. Under these conditions, the organization can either eliminate or delay a project or contract, with outside services provided to assist in completing all deliverables on schedule. These outside services may become involved in design work, programming, testing, auditing, and other staff functions.

Another type of peak load situation involves a project of limited duration. For economic reasons, many companies conclude that the use of consultants will reduce actual cost. In the current business environment, hiring permanent employees represents many expenses in addition to salary (e.g., placement fees, benefit plans, administrative costs, and training and orientation). Moreover, work sufficient to justify the additional personnel may not exist upon project completion. The resulting frequent hiring and layoffs can destroy an employer's reputation with prospective DP staff, thus making the long-range cost to the firm incalculable.

Special Skills. Organizations often find themselves requiring background or knowledge that is not readily available from internal staff. Some of these skills may be esoteric and thus unnecessary on a regular basis. Others may be quite common but, because of the goals of the organization, not available in-house. An engineering firm, for example, may not have anyone capable of generating financial systems. The entire staff may be technically oriented and well trained but not versed in the accounting side of business.

Certain management functions may also require social skills. Organizational structure, long-range planning, training, employee evaluations, staff searches, hardware/software selection, special studies, project planning and management, and many other capabilities fall into this category.

Objectivity. There are situations in which an outsider's objectivity is required—when someone is needed to rise above company politics and offer an independent viewpoint. Often the consultant is used as a buffer between competing factions within the organization. The presence of an outsider can assist these groups in resolving conflicting goals in the best interests of the organization.

In some cases, a consultant is hired to review procedures that were followed to reach a given conclusion. The assignment is not to redo the work but to verify that accepted or defensible practices were used. The consultant acts as the seal of approval.

Many times a single consulting assignment combines parts of these three areas. For example, it may be necessary to plan for and select new equipment. This requires a special skill as well as objectivity, and the consultant can provide the disinterested third-party view as well as specialized knowledge.

ESTABLISHING AND DEFINING DELIVERABLES

It is not enough to have identified a proper reason for bringing in a consultant. To successfully use one, an organization must be willing and able to properly define the results expected. Without such preliminary definition, no consulting engagement can hope to reach a satisfactory conclusion. A client who has not properly done his or her homework should entertain a proposal from the consultant to help define expected project results.

A proven, effective method for measuring progress toward goals is through the definition of deliverable results. Initially, broadly defined goals can assist in establishing project direction; however, this will be inadequate for project control and quality assurance. Deliverables must be defined in detail. It is not enough simply to specify that a report be generated. A detailed outline of that report is recommended. Major sections of the report, in fact, can become deliverables that provide client and consultant with an ability to measure progress. This procedure also provides review documents that enable the client to constantly monitor the efforts of the consultant and verify that the proper direction is being followed.

Well-defined, measurable deliverables provide a means of good project control. More important, they eliminate guesswork when identifying progress in the project life cycle. A detailed map should be available so that all parties can measure and understand the status of a project.

Consulting can be divided into two major areas that produce two types of deliverables. First are the contracting firms that provide services related to the implementation of programs, documentation, and turnkey systems. Deliverables can include programs, results of unit or system tests, documentation, hardware installation, and demonstrations.

Second are the advisory services. Deliverables in this case may not be as easy to define. They usually fall into the categories of reports and presentations encompassing anything from reorganization to training. Outlines and definitions of each section can provide the detail and intermediate milestones required.

Deliverables, thus, can take many forms. A client may require a systems design or program modification. The deliverable may be a plan for improving management control or even for deciding what the problem really is. It may simply be the availability of an independent party to review various ideas.

Ongoing support, plan reviews, and assistance in hiring staff can all be deliverables.

TIMING AND COST CONSTRAINTS

Consultants are not miracle workers. Even the very best cannot provide results overnight. This should be kept in mind when establishing contract goals. It is always in the client's best interest to ensure the environment is conducive to successful project work.

A major consideration is timing. Many clients do not contact a consultant until it is too late to complete the project properly. They then expect someone else to make up for their lack of advance planning or to accept the blame for delays.

The amount of time necessary to do a job correctly must be allowed. There is an old saying, "If you don't have time to do it right, where will you get time (or money) to do it over?" The client and the consultant must be aware of all critical deadlines. These generally concern government regulation or major milestones in corporate development (e.g., as the start-up of a new plant or the ability to respond to a new marketing plan that is already being put into effect).

It is not sufficient to look only at required dates. There is also a need to review staff availability. A consultant cannot know more about the organization than those who work there. If there is no time for interaction with the affected employees, results cannot be guaranteed.

It may be in the organization's best interest to provide the consultant with employees. This can be very advantageous to the organization, since the consultant can share his or her knowledge with in-house staff. The company may be able to gradually develop its own resources to minimize future requirements for outside assistance.

In some cases, the consultant will require additional people who are not available internally. The consultant must be able to complete the project on time and to commit additional professional staff if necessary, although five people will not necessarily complete a project in 20 percent of the time originally bid for one person. If timing is critical to the organization, however, the extra cost in overhead may be justified.

The total cost of using a consultant should always be considered. Excessive concern with hourly rates is nonproductive and can even become an obsession detrimental to the project as a whole. A $50 per hour rate may be more cost-effective than a $40 rate for reasons of speed, experience, or other factors affecting project completion. Low-ball bids have other problems as well. If the consultants underbid because they were "hungry," they might lose interest if a new project comes along at a better rate. Someone working for below-average wages will not be the best performer. Decisions should be based on dollars for results. If an emergency project cannot be cost-justified, it is probably not that important.

Finally, prerequisites should be defined. Necessary internal approvals should be known in advance. Information requirements should be defined to enable the consultant to assist in structuring a proposal that helps meet the organization's goals. Time and cost constraints on the consultant and the client should be documented. Consultants have only one resource to sell—time. If the consultant knows that the client recognizes and respects this, he or she may be willing to assist in advanced planning before hourly billing begins.

LOCATING CONSULTANTS

Once the project is defined, it can be used as a basis for determining which type of consultant would be most helpful. Consideration of the following elements will aid an educated search for individuals or firms who can fulfill an organization's needs.

Large versus Small Firm

The first element to be examined is whether a large or a small firm should be employed. Depending on circumstances, each has advantages and disadvantages. Both can provide specialized expertise and/or temporary staff to solve a client's problems. Both are able to expand the capabilities of the in-house staff. Each can provide expert opinions and an independent view.

Small Firms. The small firm has a significant advantage for smaller contracts (i.e., any project whose total cost will be less than $50,000). A large firm may be tempted to use a contract like this as a training project for new employees. Such a contract may, however, represent a significant portion of a smaller consulting firm's yearly gross. The project will thus be afforded the attention and consideration the client feels it must receive. The project will be staffed by senior or management-level people; the consulting team will probably include an owner or a director of the firm and, therefore, will receive the attention and commitment that is the basis for successful consulting.

A small firm may also be less expensive. Lower overhead and less nonproductive administrative time enable the smaller consulting firm to provide high-quality services at a lower cost.

Finally, a small firm can be very flexible. Changes in contracts and requirements can be handled and approved quickly. It is unnecessary to fight multiple levels of authority to effect minor changes.

Large Firms. The large firm has a significant advantage when handling very large projects. Several small firms can enter into a joint venture to provide for the large contract, but the large firm can respond to the same requirement and fit it into existing structure and standard project procedures. In addition, the large firm, because of its size, may be able to provide such support services as data entry, machine time, and other clerical as well as DP functions.

The large firm is more highly structured. This may help standardize and control the work process, which carries with it a risk of standardized solutions; however, an astute client can ensure that this does not happen.

Size can provide a false sense of security, however. Even the largest firms cannot justify great depth in every specialized skill. Size alone, therefore, does not provide a significant advantage except possibly when related to large projects.

One other consideration is important: whether the client feels more at ease with an officer of a small firm than with a manager of a large firm. Teamwork is basic to good consulting, and interpersonal relationships are the foundation of a good consulting environment.

General versus Specialized Consultants

The second element to consider in locating consultants is whether to contract with a firm having a broad or a specialized background. The general consultant is one who has been involved in many projects for several different industries. The other alternative is an individual who is specialized in one industry, process, language, or machine.

Many general consultants consider lack of experience in a certain situation to be valuable. In fact, they are careful not to let prior knowledge of a situation affect their investigation. They therefore do not assume some factors to be obvious and not in need of investigation. Even elementary questions are asked so that a true understanding of a client's situation and requirements can be developed. A diverse background has given these consultants the ability to examine situations from unusual angles. In looking for answers, they can review combinations and permutations of various elements from other projects in which they have been involved. New solutions may be found to old problems.

A general consultant is not always appropriate, however. A company might feel more comfortable with someone who has in-depth knowledge of the specific application. This feeling of security may be necessary to provide the comfort level required for project success.

The specialized consultant can also bring experiences from similar situations to bear on the problem. It is more likely that such consultants have been through the major pitfalls associated with certain kinds of solutions.

Consultants with strong specialized backgrounds can better lead management that is weak in state-of-the-art technology. Because of new technology and products, management may not be current in technology or confident in its own abilities. An experienced, specialized consultant can provide that extra measure of confidence necessary for success.

On the other hand, a general consultant and a strong, self-assured management team can explore unique solutions. The approaches examined for any situation can be quite varied and touch on the state of the art. The artificial constraints of convention can be replaced by new methods, possibly leading to the discovery of new processes with the potential to provide significant competitive advantages.

Each consulting situation is different, and there are excellent reasons for using each type of consultant. The decision on which to use must be based on the requirements of the project at hand. Consideration must also be given to the personalities of both the organization and the individuals involved.

Type of Contract

The last element to consider prior to selecting a consultant is whether to seek a fixed-cost or a time-and-material contract. Generally, open-ended contracts are based on time and material because sufficient information is unavailable to make a firm fixed-cost bid. This type of contract is also appropriate when a company is using consultants as an extension of its own staff.

Fixed-cost contracts provide the organization with the ability to evaluate projects on a business basis (i.e., on the known value of deliverables). Fixed-cost contracts, however, require in-depth knowledge of what is to be done. The company must be prepared to have or to develop detailed definitions for all deliverable items.

A third special category of contract exists, generally referred to as a retainer contract. Usually, the client pays a fixed amount for access to the consultant for a certain number of hours in a given period (e.g., monthly, bimonthly, quarterly). In return for the advance commitment, the consultant often charges a fee significantly less than published rates.

Retainer agreements take many different forms. Some items to consider are:
- What if more hours are required than are paid for?
- How long is the commitment?
- When are fee structures reviewed?
- What if the consultant is unable to perform?

Retainer contracts are signed for many reasons, including:
- Continuing assistance during implementation of a project
- Participation in long-range planning
- Evaluation of performance on a regular basis
- Regular training of staff
- Facilities management of equipment and/or people

Searching for Consultants

Having weighed the issues of large versus small firms, general versus specialized consultants, and the type of contract desired, the organization can now begin searching for its consultants. The organization has determined what is required, when it is needed, how it is to be completed, and what is desired in a consultant. These decisions form the foundation for a successful consulting relationship. The client can then communicate requirements so that the consultant can respond with a proposal.

The best source of consultant names is personal referrals. A recommendation from a respected associate is the best reference any consultant can have. In such a case, an individual's reputation is on the line, and he or she will not make such a recommendation lightly.

The national Independent Computer Consultants Organization (ICCA, PO Box 27412, St Louis, Missouri 63141) and its local chapters provide lists of consulting organizations. It is important to note that these are referrals rather than recommendations. The contracting firm must verify that the consultant can properly complete the project.

Finally, there are the yellow pages. Headings to check include Data Processing Services, Computer Programming Services, and Data Systems—Consultants and Designers.

A Request for Proposal (RFP) can be distributed to all potential consultants. The grapevine will carry it to firms that would not be found otherwise. (The complete details of an RFP are beyond the scope of this chapter. The major element is a repetition of the data gathering described earlier.)

Final selection requires evaluating every alternative. An easy answer to the question of how many consultants should be considered does not exist. Too many can be confusing; too few may not provide an adequate choice. The important point is to search until the right consultant is found.

SELECTING CONSULTANTS

Selecting the best consultant is as important, if not more so, than hiring an employee; a resource is being obtained from whom immediate results are expected and needed. The organization must find someone with the skills and experience to provide such assistance.

The first task in investigating consultants is to contact references. It is unlikely that a consultant would give a bad reference; therefore, questions such as "Did they get the project done?" are worthless. Questions should be designed to discover how the consultant worked; for example, asking about milestone reporting, presentations, teamwork, and interpersonal skills—things that can spell the difference between success and failure—provides valuable information. It is also important to determine the "personalities" of the companies at which particular consultants have been successful. Consultants who work well in a structured organization may have difficulty in another environment.

Interviewing Consultants

After a list of suitable consultants has been made, each must be interviewed. Feeling comfortable with the person or group is critical. Good consulting depends on teamwork, and a personality clash can drastically reduce the chances of success.

The total project should be reviewed with the consultant during the interview. Consultants may refuse a job because:

- They do not feel capable of completing it competently.
- They do not think there is a good personality match with the company.
- Time and money constraints may be such that chances of success seem low.
- They are not interested in the proposed project.

These are things to discover early. Any other concerns the organization has should be covered, including timing, additional personnel, or cost. No subject that affects the success of the project is taboo.

Cost should not be made the all-important topic of the interview, especially periodic rates. A lower hourly charge will not necessarily result in a lower project cost. Most charges are based on three factors: length of contract, individual background, and skill requirements. Length of contract is an easy measure to understand. The longer the contract, the less time the consultant must spend marketing himself or herself for future engagements in relation to the number of hours worked. That cost can be spread over a greater period and result in a lower cost per hour or day.

The individual's background is also important. A consultant with a PhD and 20 years of experience charges more than a recent programming institute graduate.

The cost of hiring a highly skilled consultant may be tempered by the skill required. For example, recommending a hardware/software solution costing more than $5 million is more expensive than designing a name and address file on a small business computer.

Consultants should submit a written proposal that should contain sufficient information for evaluating an approach and developing some idea of total cost. The client can provide the consultant with an outline specifying what is to be included in the proposal. As mentioned previously, a consultant has only one commodity—time. If the consulting firm is good (and busy), a response requiring excessive detail can be counterproductive. The best firms may not respond because of the expense involved.

A number of organizations have found it worthwhile to make a preliminary selection. They then establish a short, low-cost phase during which the consultant is asked to develop full, detailed plans. Dollar exposure is thus kept to a minimum, but a commitment has been made on both sides. A clear understanding that additional project work is dependent on satisfactory results is important if this approach is taken.

CONTRACTS

Contracts are often regarded with terror. Some organizations spend hundreds of thousands of dollars on fine print that often confuses more than helps. At the other extreme are those who feel that business on a handshake is all they need. There are even some who believe the law to be well enough defined that they are safe no matter what they sign.

The preferred view is that contracts are not the basis for suit but the basis for understanding. A contract, clearly stated, can establish an enforceable agreement that is understood and approved by both sides. The following discussion is based on the ICCA's Standard Form Consulting Contract (see Appendix), used here with permission of that association.

It must also be remembered that, when necessary, details of the contract can be changed. In such cases the basic contract discussed here may suffice, but modifications should be added to document any other understandings reached. With the proper attachments, this contract can be used for fixed-cost, time and material, and retainer contracts.

The first paragraph in this contract, entitled Services, is the most important. It goes a long way toward ensuring good consulting. A complete definition of what is to be done must be attached and signed by the principals involved. This paragraph also guarantees a consultant's ready access to the client's staff and resources as necessary. A consultant cannot perform duties adequately in a vacuum.

The second paragraph is entitled Rate of Payment for Services. Nothing is left to speculation. Everyone involved must state on paper what is expected, when, and for what cost. This will eliminate almost all arguments usually associated with contracts that seem to have gone sour.

Paragraph 3 is related to expenses. The wording should be based on the organization's situation. Specific reimbursable or nonreimbursable expenses should be defined before work begins.

Paragraph 4 is a simple statement that the client will pay the amounts agreed to in paragraphs 2 and 3 upon receipt of invoices. This forces the consultant to follow standard business billing practices.

Paragraph 5 covers confidential information. Both the client and consultant are protected. They both agree not to disclose to an outside party any confidential information on research, development, trade secrets, or business affairs. This, of course, refers to information not generally known or "easily ascertainable by non-parties or ordinary skill in computer design and programming."

Paragraph 6 is designed to help both client and consultant protect their personnel resources. Both parties agree not to try to hire the other's employees for at least six months after project completion, except by written agreement. Because of a number of legal cases that have arisen recently, it further states that "neither consultant nor consultant's staff is or shall be deemed to be employees of the client." Consultant staff may include full-time employees and/or subcontractors.

Paragraph 7 defines ownership of deliverables produced during the project. This paragraph is frequently changed. In some cases, a client will make special arrangements with the consultant to receive pricing considerations in return for releasing ownership and future marketing rights. These questions should be worked out well in advance and worded clearly so that each party's rights and privileges are understood.

Paragraph 8 is one of the most important to the consultant. The client liaison, responsible for, and with the authority to control, the project, is listed by name.

Paragraph 9 concerns warranties and consultant liability. This is the most legally complicated paragraph in the contract. Included is an agreement to attach to the contract any special requirements for formats or standards to be followed in the project. That is followed by a statement concerning warranties, "whether written, oral, or implied," that follow specific legal standards.

Paragraph 10 is a simple legal statement that specifies that this document is the complete agreement.

Paragraph 11 identifies the state law under which the contract is to be signed. It is generally the consultant's home state; however, many clients alter this to their own state if different from that of the consultant. It is a minor point (unless the client expects the contract to go to suit) and should be negotiated between parties.

The twelfth paragraph, entitled Scope of Agreement, is a way of legally covering all bases.

Paragraph 13, entitled Additional Work, outlines the procedure to be followed when the services requested are changed or added to. The process may be modified to fit the client's standards. The usual minimum requirement is that the client submit a written request for additional services.

Paragraph 14 identifies the official addresses of both client and consultant.

Paragraph 15, the last standard paragraph, is a legal formula prohibiting assignment by either party without the prior written consent of the other. The parties agree to complete the contract.

Additional clauses can be added as necessary. The goal of the contract is to define what, when, and how much is involved. Any special needs, agreements, or arrangements should thus be spelled out.

An alternative to this contract is possible. A simple "letter of understanding" that identifies the services, payments, ownership of the final product, and the client representative may suffice. There is no absolute requirement for legal format. The most important element is that the parties have documented and reviewed their agreement with each other.

CONCLUSION

Consultants offer management the chance to expand the abilities of their organization and are a valuable resource to be sought out and used. The possibilities can be endless. Nonetheless, successful consulting requires a team effort. Management commitment makes the difference between success and failure.

APPENDIX

Independent Computer Consultants Association

STANDARD FORM CONSULTING CONTRACT

THIS AGREEMENT is made as of _____, 19 _____ .

between _____ ("Client")

and _____ ("Consultant")

In the event of a conflict in the provisions of any attachments hereto and the provisions set forth in this Agreement, the provisions of such attachments shall govern.

1. **Services.** Consultant agrees to perform for Client the services listed in the Scope of Services section in Exhibit A, attached hereto and executed by both Client and Consultant. Such services are hereinafter referred to as "Services". Client agrees that consultant shall have ready access to client's staff and resources as necessary to perform the Consultant's services provided for by this contract.

2. **Rate of Payment for Services.** Client agrees to pay Consultant for Services in accordance with the schedule contained in Exhibit B attached hereto and executed by both Client and Consultant.

3. **Reimbursement for Expenses.** Consultant shall be reimbursed by Client for all reasonable expenses incurred by Consultant in the performance of Services, including, but not limited to, travel expenses of Consultant and Consultant's staff, long distance telephone calls, computer time, and supplies.

4. **Invoicing.** Client shall pay the amounts agreed to herein upon receipt of invoices which shall be sent by, and client shall pay the amount of such invoices to Consultant.

5. **Confidential Information.** Each party hereto ("Such Party") shall hold in trust for the other party hereto ("Such Other Party"), and shall not disclose to any nonparty to the Agreement, any confidential information of Such Other Party. Confidential information is information which relates to Such Other Party's research, development, trade secrets or business affairs, but does not include information which is generally known or easily ascertainable by non-parties of ordinary skill in computer design and programming.

6. **Staff.** Neither Consultant nor Consultant's staff is or shall be deemed to be employees of Client. Consultant shall take appropriate measures to insure that its staff who perform Services are competent to do so and that they do not breach Section 5 hereof.

Each of the parties hereto agrees that, while performing Services under this Agreement, and for a period of six (6) months following the termination of this Agreement, neither party will, except with the other party's prior written approval, solicit or offer employment to the other party's employees or staff engaged in any efforts under this Agreement.

7. **Use of Work Product.** Consultant and Client agree that Client shall have nonexclusive ownership of the deliverable products described in Exhibit A and the ideas embodied therein.

8. **Client Representative.** The following individual _____

_____ shall represent the client during the performance of this contract with respect to the services and deliverables as defined herein and has authority to execute written modifications or additions to this contract as defined in section 13.

LIMITED WARRANTY

9. **Liability.** Consultant warrants to Client that the material, analysis, data, programs and services to be delivered or rendered hereunder, will be of the kind and quality designated and will be performed by qualified personnel. Special requirements for format or standards to be followed shall be attached as an additional Exhibit and executed by both Client and Consultant. **Consultant makes no other warranties, whether written, oral or implied, including without limitation warranty of fitness for purpose or merchantability.** In no event shall Consultant be liable for special or consequential damages, either in contract or tort, whether or not the possibility of such damages has been disclosed to Consultant in advance or could have been reasonably foreseen by Consultant, and in the event this limitation of damages is held unenforceable, then the parties agree that by reason of the difficulty in foreseeing possible damages all liability to client shall be limited to One Hundred dollars ($100.00) as liquidated damages and not as penalty.

10. **Complete Agreement.** This agreement contains the entire agreement between the parties hereto with respect to the matters covered herein. No other agreements, representations, warranties or other matters, oral or written, purportedly agreed to or represented by or on behalf of Consultant by any of its employees or agents, or contained in any sales materials or brochures, shall be deemed to bind the parties hereto with respect to the subject matter hereof. Client acknowledges that it is entering into this Agreement solely on the basis of the representations contained herein.

11. **Applicable Law.** Consultant shall comply with all applicable laws in performing Services but shall be held harmless for violation of any governmental procurement regulation to which it may be subject but to which reference is not made in Exhibit A. This Agreement shall be construed in accordance with the laws of the State indicated by the consultant's address (14ii).

12. **Scope of Agreement.** If the scope of any of the provisions of the Agreement is too broad in any respect whatsoever to permit enforcement to its full extent, then such provisions shall be enforced to the maximum extent permitted by law, and the parties hereto consent and agree that such scope may be judicially modified accordingly and that the whole of such provisions of this Agreement shall not thereby fail, but that the scope of such provisions shall be curtailed only to the extent necessary to conform to law.

13. **Additional Work.** After receipt of an order which adds to the Services, Consultant may, at its discretion, take reasonable action and expend reasonable amounts of time and money based on such order. Client agrees to pay and reimburse Consultant for such action and expenditure as set forth in Exhibit B of this Agreement for payments and reimbursements related to Services.

14. **Notices.**

(i)   Notices to Client should be sent to:

_____

_____

(ii)   Notices to Consultant should be sent to:

_____

_____

15. **Assignment.** This Agreement may not be assigned by either party without the prior written consent of the other party. Except for the prohibition on assignment contained in the preceding sentence, this Agreement shall be binding upon and inure to the benefit of the heirs, successors and assigns of the parties hereto.

IN WITNESS WHEREOF, the parties hereto have signed this Agreement as of the date first above written.

_____   _____

Client   Consultant

_____

type Name and Title

5 Systems Analysis Checklist

by Raymond P. Wenig

INTRODUCTION

A systems analysis checklist can improve the results of the analysis and the overall effectiveness of the project team. It can also help produce consistent results and contribute to the expertise of the team members who perform the analysis. This chapter presents a checklist for planning and evaluating the systems analysis phase of a project.

CHECKLIST CONTENTS

The following questions cover the major areas of evaluation and review to ascertain that systems analysis work is progressing steadily.

Analysis Planning

Questions
1. Are the reasons for the analysis project clearly defined in writing?
2. Are the project limits defined?
3. Are limits set on resources, time, and funds?
4. Is completion of the system scheduled?
5. Who will perform the analysis work?
6. Who are the user participants?
7. Are objectives set for the new or modified system? If so, what are they, and who set them?
8. What priority has the organization set for the project?
9. What previous systems analysis work has been performed in this application area?
10. What is the status of current systems serving the application?
11. What (if any) special legal, security, or audit considerations must be observed in this system?

Output
1. A narrative definition of the project boundaries
2. A tentative work plan for the analysis work
3. A user contact list

4. A tentative resource staffing list
5. A list of existing application systems
6. A priority impact statement concerning the relative importance of the system

User Contacts

Questions
1. Are all user participants identified?
2. What are the organizational relationships of the users?
3. What is the current level of user systems knowledge? Have the users had previous systems experience?
4. Do users clearly understand the current system and its operation?
5. Are legitimate user complaints about the current system documented? Is the impact of the complaints fully documented?
6. How much time and effort are the users willing to put into the initial analysis work?
7. Are users identified who are supporters of, resistant to, and indifferent to the system?
8. Do users expect any specific benefits from the resulting system?
9. Is there clearly defined top-level support for the project? If so, who constitutes this support, and how much power do they wield?
10. Who are the key decision makers in the user environment?
11. How many user locations are there? How many people will use the system at various levels? What is their level of computer systems experience?

Output
1. An organizational chart of all participating user areas, including their hierarchical relationships
2. A narrative describing the users' systems backgrounds and prior experiences
3. Documentation of user problems with the existing system and the impact of these problems
4. A work plan of expected user participation in the analysis
5. A tentative statement of user expectations
6. A narrative on the political relationships and systems support expectations of the major user participants
7. A brief history of previous data systems and procedures used in the application area
8. Identification of any other organizational systems or applications that interrelate with the proposed system

System Objectives

Questions
1. Are system objectives formally defined, or are they loosely stated and subject to interpretation and/or later definition?

2. Will the new system have a major impact on the basic operations of the organization?
3. Will the new system replace an existing one? If so, how old is the current system, and how many others preceded it?
4. Is the new system expected to cause relocation or removal of any work functions? If so, how sensitive is the issue, and who will help combat any resistance?
5. Is an interim system required to satisfy immediate goals or to eliminate intolerable problems with the existing system?
6. Is a phased development and implementation approach feasible, or is a one-time mass conversion required?
7. What cost can be justified, and what resources can be allocated for this project?
8. How close to the state of the art is the new system expected to be?
9. How much organizational shock can users tolerate? How much change do they really want? How much change will cause them to reject the new system?
10. How much time can users allocate for training and start-up? During what period of time?

Output
1. A comprehensive statement of system objectives
2. A statement of general scope and level of project effort required, including tentative cost and resource estimates
3. A statement concerning the current system and procedures considered for change, elimination, and/or replacement
4. A general statement covering the expected project phasing and the overall team approach to the project
5. A tentative statement covering the levels and impact of anticipated organizational changes that will result from the system
6. A commentary on the roles and responsibilities of each participating user department and major user group in the desired system

Current System

Questions
1. What are the problems with the current system as evaluated by the users and by the technical team? Do these evaluations agree?
2. How do other organizations perform similar functions? What is the current state of the art in the application area?
3. What other methods and procedures have been tried and/or used to service the application?
4. What is the detailed chronology of the current system, its predecessor, and the changes made to both systems?
5. What is the organization's history during the current system's life?
6. What development, maintenance, and operational costs are associated with the current system (including user efforts)?

7. Identify the name, rank, and organizational position of those who supported, built, and use the current system.
8. Identify one or more major situational failures that resulted from the current system.

Output
1. A comprehensive narrative on the current system and its operation, history, and users
2. A ranked list of the current system's major faults and problems
3. A full cost analysis of the current system
4. A general statement on how closely the new system might approach those in other organizations or the state of the art
5. A complete collection of the documents, procedures, and other available details concerning the operation/content of the current system

Data Elements and Structures

Questions
1. Are the current data elements, files, forms, procedures, and so on thoroughly documented?
2. Are the current data elements and structures logical, consistent, and utilized?
3. How clean is the current data base?
4. Do users have a list of new data elements they would like to see in the new system? Is it feasible to add these data elements?
5. How much redundancy exists between the current system's data base and that of other applications in the organization? Are any of the other applications a more logical repository for any elements of the data base?
6. Is there enough flexibility in the current data structure to perform efficient modifications or changes? Can the structure be changed to meet the new system's needs?
7. How difficult will it be to convert the current data base to a new one? How much error testing will have to be done to achieve a clean conversion?
8. How much maintenance is usually done on the existing data base?
9. Can or should extensive data archives from this data base be converted?
10. How much of the current data base is actively used? By whom?
11. What significant faults or failures were encountered with the data files, and how were they dealt with?
12. How many times and in what ways has the data base been modified?

Output
1. A comprehensive set of format and content definitions of all data elements, files, and supporting data structures
2. An evaluation of current data base content, with emphasis on cleanli-

ness, errors, unused areas, redundancy, conversion, and future use
3. A list of expected changes, additions, deletions, and other modifications to data elements and structure that are anticipated for the new system
4. A summary of the major uses of the data file and its elements
5. A list of faults and failures of the existing data files

User Interviews

Questions
1. Are all users identified?
2. Is there a formal interview plan for each user level covered?
3. Are lists of questions and objectives developed for the interviews at each user level?
4. Is top management supporting and publicizing the interviews, the interview team, and the overall expectations? Is top management making a strong pitch for interviewee cooperation?
5. Are all interviews scheduled during acceptable time periods?
6. Are the interviewers trained in effective interview techniques?
7. Are all scheduled interviews completed? Have canceled, interrupted, or forgotten interviews been rescheduled and conducted?
8. Have the interviewers taken adequate notes and written evaluations of each interview?
9. Have the interviewers compared notes, impressions, and other observations? Are these details documented?
10. Are interviewees given adequate feedback (e.g., summary reports, notes)?
11. Have follow-up interviews been conducted when special problems or conditions are uncovered during initial interviews?
12. Has management been kept informed about the interview process, any problems uncovered, and uncooperative users?

Output
1. A formal interview plan
2. Documentation of interview results
3. A report summarizing the interviews that includes both consensus answers and significant variances
4. An internal analysis of user attitudes and positions vis-a-vis the system
5. A management report covering interview findings and cooperation of the participants
6. Results of test interviews along with changes in questions, emphasis, and other interviewing guidelines
7. Explanation of any incomplete interviews

Research on Other Systems

Questions
1. What other organizations can be surveyed regarding their approach to the subject application?

2. What (if any) proprietary packages are available that might suit the application area?
3. What (if any) trade/industry associations study or catalog the systems work of others in the same field?
4. What (if any) formal literature is available on the subject application area?
5. How much time and effort should be spent in reviewing other systems?
6. Were the reviews of other systems productive? Should more time be spent on this activity?
7. Are field interviews of other users and organizations necessary?

Output
1. A list of organizations and sources to review for base knowledge on alternative approaches to the application
2. A narrative report detailing the ways other organizations are solving the application
3. A technical evaluation covering the state of the art for the application area
4. A summary report on contacts with other users and organizations
5. A follow-up plan for reviewing or tracking major developments in the industry

Alternative Propositions

Questions
1. How many application alternatives should be considered?
2. How much time and effort should be spent in evaluation of alternatives?
3. How detailed and complete should the consideration of each alternative be?
4. How will the alternatives be developed and documented?
5. Are formal requirements and evaluation criteria established for the alternatives?
6. Who will evaluate the alternatives? Will the users review the alternatives?
7. Are all logical alternatives being considered?
8. Are outside expert opinions being sought on the alternatives?
9. Are the alternatives considered consistent with those evaluated by other organizations?

Output
1. Alternative design definitions
2. Positive and negative factors of each alternative
3. Evaluation reports from each group who studies the alternatives
4. Formal user presentation of the alternatives
5. Preliminary cost predictions for each alternative

6. A technology impact assessment for each alternative
7. A user impact assessment for each alternative

Selecting a Design Alternative

Questions
1. Are all alternatives fully reviewed and evaluated?
2. Are the alternatives ranked in terms of their ability to meet the system requirements criteria?
3. Is there a technical/management team with authority to select the most appropriate alternative?
4. Does one alternative clearly outrank the others?
5. Which alternative(s) do the users support?
6. Which alternative is best to implement in terms of time, cost, resources, and technical risk?
7. Which alternative uses the most advanced concepts?
8. Which alternative is likely to last the longest?

Output
1. A detailed comparison of alternatives
2. A ranking of alternatives
3. A specific recommendation as to the alternative that is best to pursue
4. A report to the users on the alternative selected
5. A summary of reasons for rejecting other alternatives

Structural Analysis

Questions
1. Are all data elements, flows, and expected processing steps defined for the selected alternative?
2. Are procedural and organizational changes that the new system will generate defined and evaluated?
3. Are the content and uses of input files and outputs defined in a general way?
4. Are the equipment requirements for the new system estimated?
5. Is there a list of expected system modules?
6. Is there a tentative data conversion plan?
7. Is an overall system flow being generated?
8. Are associated clerical procedures outlined?
9. What is the estimated volume of data and transactions?
10. Are the security and accuracy requirements of the data being considered?
11. Are testing procedures for the new approach thoroughly defined?
12. Is a preliminary system implementation plan available?

Output
1. A report of the proposed system approach
2. A system flowchart

3. A user operations and responsibility flowchart
4. A detailed report on the analysis findings
5. A cost/benefit analysis report
6. A preliminary testing plan
7. A tentative implementation plan

Plans for Next Phase

Questions
1. Are there work tasks and resource estimates for the general design work?
2. Is there a resource loading plan that shows requirements by work task?
3. Are user support tasks identified and planned? Are the users aware of them?
4. Are target dates set for obtaining authorization to proceed with the next phase? What is the expected completion date of the proposed work?

Output
1. The work plan and resource estimates
2. The user support plan
3. A narrative on the approach to managing the next phase

Management Presentations and Reviews

Questions
1. Are all levels of management in the technical and user areas briefed on the analysis results and recommendations?
2. Are the presentations clearly and logically formulated?
3. Are management's concerns and questions documented and answered?
4. Has the proposed alternative survived management's scrutiny?
5. Does the analysis team have any doubts about the project approach?
6. Have minority opinions and negative comments been properly addressed?

Output
1. Presentation critiques and internal reviews
2. Presentation reports and visual aids
3. Authorization to proceed

CONCLUSION

A checklist can expedite and help ensure the high quality and completeness of systems analysis work. The checklist presented in this chapter can be used as is or can be modified to suit the organization, the users, or the specific projects.

⑥ User-Oriented Systems Analysis and Design

by Henry C. Lucas, Jr.

INTRODUCTION

Users often complain about the small return from their large investment in computer-based information systems. They are frustrated by their inability to influence decisions about information systems in their organizations. Often, expensive computer-based systems are not used at all or are not exploited to their full potential. Experiences in different organizations have produced the following examples of problems:

- Two information systems at a major bank calculated the internal transfer price for borrowing and lending among branches. Each system's report showed a different figure, which clearly should have been identical on each output. Branch managers questioned both figures and did not rely on any of the data in the two reports because of this inconsistency.
- The manager of the computer department in a manufacturing company had not distributed computer output for two months because he was not completely satisfied with the reports. Users did not seem to notice the absence of the output.
- A major university developed a sophisticated online computer system to automate a number of administrative functions. Most users expressed a desire to return to manual or batch computer-produced reports because of difficulties with the new system.
- One mining company spent almost five years designing an inventory control system for its largest division. When installed, the system showed clear cost savings. Several years later, however, managers in other divisions were still resisting the installation of the new system.
- A manufacturing company installed a modified order-entry package. The system eventually worked, but during installation the company lost track of all orders for three days.

What is responsible for these problems with information systems? Why are systems analysts and systems designers creating systems that are not used? This chapter suggests a method of systems analysis and design centered on the user. The theory behind user-oriented design is that the systems analysis and design phase is too important to be left solely to the professional designer. In user-oriented design, responsibility for the system shifts from the analyst to the user.

TYPES OF INFORMATION SYSTEMS

Before discussing some of the details of user-oriented design, it is useful to describe different information systems and to review the systems life cycle. Different individuals in an organization make different decisions. Rarely is the lowest supervisory level in a company involved in strategic planning decisions. On the other hand, the president of an organization makes relatively few operational control decisions. Thus, when a system is being designed for a particular level of management, an analyst should keep in mind the type of information required. Information systems requirements fall into three categories defined by the types of decisions they support [1].

Strategic planning decisions determine the objectives of an organization and allocate resources to attain these goals. These decisions are made over a long period of time and often involve substantial investment. The development and marketing of a new product is an example of a strategic decision, as is the commitment to acquire a new subsidiary.

Managerial control decisions are concerned with the use of resources in the organization. These decisions often deal with financial or personnel considerations. An accountant trying to determine the reasons for a deviation from planned budget is working on a managerial control problem.

Operational control decisions deal with the daily operations of the firm and tend to be short run in nature. What the factory should produce today and how much of a certain part should be reordered for inventory are operational control questions.

Information for strategic planning (e.g., data on the economy, competition) usually comes from the external environment. Accurate detailed information is not mandatory for strategic planning; summary information may be all that is needed in many situations. Strategic decisions usually involve planning and are more long-range than other decisions.

Operational control decisions have almost opposite information requirements—the data for operational control decisions usually is generated internally, and accuracy is highly important. Detailed information is the rule, and this type of decision must be made frequently. Operational control decisions are of short range and are likely to trigger immediate action. The information requirements for managerial control decisions fall in between those of the other two types.

A type of information system that cuts across the categories described above is the decision support system (DSS). DSSs are designed to support a specific decision, like those made in portfolio management and production planning. DSSs often involve mathematical models and large data bases. One of their main characteristics is voluntary use.

The typical production system (e.g., one that processes payments, orders, and shipments) must be used; a DSS, on the other hand, is adopted voluntarily by the decision maker. There are numerous sophisticated and relevant systems that are not used by those for whom they were designed. One company with a large decision support system estimates that only 10 percent of the potential users actually use the system.

The Designer's Responsibility

The designer should recognize the different information requirements of the various types of decisions. One of the largest problems in the design of information systems (especially top-management decision systems) is that of providing the wrong data for a particular decision. Analysts, conditioned by lower-level operational control systems, may supply top management—faced with a strategic problem—with unnecessary data generated from internal records with high levels of accuracy and detail.

Strategic planning, managerial control, and operational control can be supported by computer-based information systems. Most current computer-based systems, however, are transaction-oriented systems, which involve very few decisions or decisions that are so routine and programmed that they are uninteresting. For example, they compute the payroll or produce accounts payable checks. Frequently, however, transaction-processing systems collect the information necessary to make other kinds of decisions. An order-entry system, for example, may produce summary reports that are useful in solving operational control problems, such as production scheduling.

There is nothing wrong with developing transaction-processing systems— they are often able to demonstrate cost savings to the organization. The organization that develops only this type of system, however, ignores some of the potential of the computer as an aid to decision making. Good transaction-processing systems are necessary, but they should not be the only types of computer-based aids developed.

CONVENTIONAL APPROACHES TO ANALYSIS AND DESIGN

This section describes some of the conventional approaches to systems analysis and design and suggests some of the problems with them.

The Systems Life Cycle

Table 6-1 shows the stages in the systems life cycle. The need to improve existing information processing procedures usually stimulates the desire for a new computer-based information system. A feasibility study or a preliminary survey is conducted to determine if a system can be developed to solve the users' information processing problems. Based on the outcome of a feasibility study, a decision is made to proceed with the design of a system.

The design stage is the major creative part of the systems life cycle. Detailed specifications are developed for exactly what the system is to do. Programming turns these detailed specifications into a working computer system, and testing ensures that the system works satisfactorily. Throughout the programming and testing stages, the system is documented.

Changes to the existing information processing procedures are made during conversion so that the new system can be used. During installation, the organization begins to rely on the new system. Finally, after installation is completed, the system becomes operational and is run on a routine basis.

Stages in the Design Process

Table 6-2 contains a list of the major tasks undertaken in systems analysis and design.

Motivation refers to the reason the study is being undertaken. Generally, a user has an information processing problem and feels that the computer can help in solving it. (Chances for success are much greater when the user, rather than the DP department, suggests a new system.) The analyst tries to determine the users' goals for the system and attempts to understand the existing system in terms of its performance of some or all of the required functions.

Table 6-1. Systems Life Cycle	Table 6-2. Steps in Systems Design
Inception	Motivation
Feasibility Study	Feasibility Study
Design	Systems Analysis
Specifications	An Ideal System
Programming	Detailed Specifications
Testing	Conversion and Installation
Documentation	
Conversion	
Installation	
Operation	

Based on initial discussions with users, a feasibility study is conducted. The feasibility study includes documentation of the existing information processing procedures. The design team then formulates a rough alternative system and estimates costs. At the completion of the feasibility study, a decision is made on whether or not to proceed with the system.

If the decision is positive, detailed systems analysis and design are undertaken. The approach must first be documented thoroughly through the collection of data on the volume of input and output, information flows, and decisions. Then the actual systems design begins. One way to produce a new system is to design an ideal one without cost or other constraints. When this is accomplished, the design team iterates to produce an acceptable and feasible system; for example, modifications are made to the ideal system to bring its costs within reasonable limits.

Following the completion of the outline for the system, detailed specifications are produced at the processing, logic, file design, and I/O levels. Programs are assigned to and written by programmers. Manual procedures are specified and the entire system tested, both with unit test data and logical data for the entire system.

During conversion and installation, existing information processing procedures are phased out as the new system begins working. These stages involve training users and running final tests as well as converting files and other procedures to the new system.

Problems with the Conventional Approach

The steps contained in Table 6-2 are conventional. Many texts on systems analysis and design contain similar lists of tasks. Problems in four areas— information flows and paper processing, decision making, change in the organization, and the role of the analyst—arise when this approach is used.

Information Flows and Paper Processing. The conventional approach overemphasizes information and paper flows. These processes are, in fact, independent of the development of computer systems and could just as easily apply to the development of systems and procedures involving manual tabulating equipment.

Decision Making. The stages in conventional design do not sufficiently consider decision making. Computer-based systems can potentially assist in making decisions, and systems designers should focus on this as well as on paper flows. The failure to do so, combined with the overemphasis on information and paper flows, has resulted in an overabundance of transaction systems. While these systems must function well if the organization is to continue in business, the potential of computer-based systems is not realized if systems do not also support decision making.

Change in the Organization. The conventional view of systems analysis and design obscures the fact that information systems are designed to bring about change in the organization. If users were satisfied with existing information processing procedures, there would be no reason for a new system. Of course, the degree of change varies from one system to another. Some implementation efforts involve only minimal changes, such as new input or output procedures; others may result in changes to work groups or the structure of the organization. Whatever the case, an approach to analysis and design that takes into account the problems of introducing change is needed.

Role of the Analyst. The last problem concerns the role of the systems analyst. The conventional design method implies that the analyst is completely in charge of the systems design process. The analyst is seen as an artist or an architect who receives a commission, discusses the work with the client, and creates the desired product. This has led to the failure of many systems.

USER-ORIENTED DESIGN

Rather than placing systems analysts in charge of the design effort, users should themselves manage the design of their computer-based information systems. They should actually perform some of the tasks usually carried out by the analyst. Experience indicates that users are capable of such tasks and that successful results can be achieved in this way [2].

Reasons for User Participation

There are many reasons for user participation in the design of information systems. In the past, user participation has meant that designers consulted users, but the users did not necessarily have any real influence over the system. Real user involvement requires time. Users must understand the system, and their recommendations must prevail. A number of benefits result from this type of user participation [3]:

- It builds user self-esteem.
- It is intrinsically satisfying and challenging.
- Because the users have psychological ownership of the system, they are motivated to work with it, and the new system is more likely to be used.
- More commitment to change usually results.
- Users become more knowledgeable about the system and are trained to use it prior to conversion and installation.
- Users retain much of the control over operations in their areas.
- The users know what is needed for a particular application; if the users are in charge, quality is defined according to the users' criteria.
- Users know more about present information processing procedures, and user-oriented design therefore results in better solutions to problems.
- User-designed interfaces are easier to use than those designed by systems designers.

A New Design Methodology

User-oriented design has three major components:
- User-controlled systems design
- User-defined criteria of system quality
- Special attention to the design of the interface between user and system

User-Controlled Design. User control of design may be innovative, but it is the most important component of user-oriented design. Although many DP departments stress participation and involvement, this involvement is often superficial. Users' suggestions are solicited, but users have little influence on the final system. In user-oriented design, the responsibility for the design of the system lies with the user. The computer professional becomes a catalyst who helps the user construct the system and who translates the user design into technical specifications for computer processing. User-oriented design places the user in total control of the design of the system.

The users' efforts are guided by the analyst, who indicates what tasks must be accomplished. For example, the analyst might suggest that the first task is the specification of output. Users are asked to think about the information desired and to draw a rough sketch of a needed report. The users then consider ways in which the report could be used. Trade-offs among different ways of making the information available (e.g., online inquiry or printed report) are discussed with the users. The analyst, based on his or her knowledge of computer capabilities, presents alternatives for user consideration.

Next, the users might be asked to develop a method for obtaining input for the new system. The users determine the content and form of the input after the analyst has discussed such alternatives as a terminal, mark sensing, and optical character recognition. Finally, the users are shown how computer files are developed. Working with the analyst, users define the processing logic and file structures for the system. Users should also suggest conversion and installation plans.

System Quality. The second part of user-oriented design is concerned with system quality, which should be evaluated according to user criteria rather than the criteria of the DP department or professional analyst. In the university system described earlier in this chapter, the DP department developed an online system using the latest in communications and data base technologies. Users were irritated, however, because the command language was difficult to use and because the system contained a number of errors. Users no longer had the familiar batch reports, and the system was available only for a short period during the day. Computer professionals thought this system was excellent because of its technical elegance, but the users were dissatisfied because the system was driven by technology more than by their needs.

User/System Interface. The final component of user-oriented design is the interface between the users and the system. Effort should be made to ensure the design of a high-quality interface. Input and output with which users have contact should be carefully designed; experimentation should be the rule. Users should design their own input and output forms and should have the opportunity to work with them and the proposed output devices before they become part of the system. Users should also choose the appropriate mode (i.e., batch or online) and technology for the system.

Reaction of the Systems Staff

The systems staff may fear loss of control if user-oriented design is employed. For example, one manager resisted this approach primarily because he was rewarded for finishing systems on time and within budget. He perceived that management wanted his systems staff to be cost cutters. This conception of the DP function suggests development of operational control systems, the use of which is mandatory. In such an environment, user-oriented design is difficult to implement.

Many professionals now realize that conventional design approaches have consistently resulted in failure and sometimes in disaster. Although it is not universally endorsed by systems professionals, there is growing recognition that user-oriented design, or a similar technique, will be required for success in the future.

ADOPTING USER-ORIENTED DESIGN

The following discussion presents a series of steps designed to aid in adopting user-oriented design.

Application Identification

A key activity for the organization is to identify areas where potential for computer applications is high. The problem does not concern the feasibility of an application but rather what type of system should be developed and what the priorities of different suggested applications should be.

The identification and selection of applications is a key place for the involvement of users. Users should understand why a particular application is chosen for development; often, higher management commissions a new system. The end user may not have had any input in the decision to develop a system. Management should make clear the reasons for the new application to everyone involved in the design and use of the system.

Users should also influence system boundaries; more than one alternative to the status quo should be considered. In some instances the user may choose a less complex system, omitting some functions in the interest of rapid implementation. In other circumstances, they may opt for a very elaborate and sophisticated application. Whatever alternative is selected, the user should consider a range of options and participate in the choice.

Design Committee

The use of a design committee is integral to user-oriented design. All levels of individuals affected by the system should participate in its design. It is difficult for one person to design a system—the more individuals involved in this creative task, the better the system. If there are too many individuals for all to be included on the design team, a representative from each group of users should be selected. The representatives then act as the liaison between the design team and the users. In the design of a retail data collection system, for example, certain clerks could represent all clerks who will use the system.

Appointing the Head of the Design Team

It is important that the head of the design team be a user. Otherwise, the users will not perceive that they are in control of the development process. One of the goals of user-oriented design is to ease conversion and to ensure that users have psychological ownership of the system. To achieve this goal, a user must be in charge of the design team.

The Role of Management

Management plays a key role in the development of a new system and the adoption of user-oriented design. Management must clarify the objectives of all new systems. In one company, management wanted to unify the customer services function, thereby removing customer services from two areas and creating a new department responsible for all customer contact. This unpopular change was blamed on a new order-processing system until it was made clear that top management wanted the change and that the new system would facilitate it.

Managers must also provide resources so that users can participate in design; for example, they may have to hire new employees to free user time. Management must also encourage and attend frequent review meetings to discuss the design in progress.

These reviews play an extremely important part in the user design process. Everyone involved with the system must attend. Management does not always understand that it should also be involved. Often during these meetings, policy questions arise that must have the input of higher levels of management. For example, management must participate in decisions on the allocation of products to customers. In addition, the participation of high-level managers in the design process serves as a model for others in the organization; this kind of participation is a part of management's leadership role.

User Role in Design

A continual difficulty in the design of new information processing systems is the lack of available time users can give to the design effort. New information systems are usually designed for areas where users are already overburdened; existing information processing procedures may have broken down. Managers of user areas must provide sufficient resources to enable user participation.

Role of the Professional Analyst

The professional analyst who adopts the role of catalyst in the design process is crucial to user-oriented design. Instead of being in charge of the system, the analyst should present alternatives to the users. Presenting the various stages of the systems life cycle is a way to start. The approach should be to ask users about decisions, rather than to tell them what the computer system is going to do. The analyst explains each alternative and its benefits, costs, and trade-offs and gives reasons for recommending a particular alternative. If a user can justify a request for 12 months' sales history for the current and previous years, it should be provided. Above all, the analyst should not speak of what he or she can do for the users but rather of what the computer can be programmed to do.

Specifying Goals

One helpful design approach is to begin by specifying the goals for the new system. A group meeting can be held to obtain an overview of what a system should accomplish. Next, users identify the output they would like to have from the system—not in detail but in broad terms. The inputs available to produce this output are described. From this the contents of files are developed. At this point, users meet with the analyst to determine the processing mode. Output displays or report formats are then developed in detail. Input documents or displays are refined and the file contents specified.

Progress Review

Although the design approach sounds sequential, it is not. There are a number of cycles in which progress is reviewed and refined. As the system evolves, frequent review meetings and walkthroughs are held. Several users should attempt to define each display or report. Again, the development of a system is a creative process, and the creativity of more than one individual is needed. Individuals should be encouraged to walk through their processing with the entire group.

Challenging the Design

One function of the systems analyst is to challenge the design as it develops. The analyst must check to ensure that the multiple uses of information have been considered. For an accountant, for example, data on last year's sales may be viewed as historical, whereas the market researcher might look at this data as indicative of future sales. The analyst must be sure that decision making, not just the flow of data, has been considered in the design.

Testing

The interface must be carefully tested. Users should develop their own pro forma reports and should review all input and output documents and displays carefully. Where possible, live tests should be conducted, and in an online system, the user should work with a terminal display before finalizing the system specifications.

Conversion and Installation Plan

Users can develop the conversion and installation plan. What data must be transferred to the new system and how different individuals will respond to a new system are important considerations. A foreman with 20 years of experience may react quite differently from a manager who has just joined the company. It is important for users to develop test data in order to assure themselves that the system operates according to specifications.

Post-Implementation Audit

A post-implementation audit should be conducted by the users and the systems analyst working on the project. Some questions to ask are:
- Were the stated goals achieved?
- Were costs within reason?

- Does the system function according to the desires of the users?
- What can be done to improve the design approach in the future?
- Were enough meetings held?
- Did users on the team understand what they were requesting

CASE STUDIES

This section presents two examples of user-oriented design. The first example involves a firm that had followed the conventional approach to design and encountered difficulty. A user-oriented approach was adopted to rescue the project.

Order Processing

This firm had developed one of the early online order-entry systems in its industry. The competition, however, had since developed more advanced systems, and this company wished to develop the "next generation." Design work had begun, and the analysts felt they were very user oriented; however, management, after receiving an 8-inch-thick set of preliminary systems specifications that it could not understand, had misgivings. A consultant was retained to evaluate user reaction to the system.

The new system was to be quite comprehensive; it was to begin with a new and more sophisticated forecasting technique, encompass order entry and production scheduling, and eventually attempt to load machinery on the production floor. A number of new features required extensive research.

The consultant confirmed management's fears—very few users really understood the system, and most had misgivings about how it would work in their environment. The consultant recommended that users and top management attend a series of review meetings.

The consultant learned enough about the system to make a presentation in the first meeting. The discussion was at the conceptual level. Users and managers from all functional areas began to understand the implications of the system and its boundaries. They reviewed the list of remaining conceptual design questions and added to it a number of further issues to be explored.

At a second meeting a month later, the remaining design issues were discussed. At the end of the meeting, the issues were grouped into categories, and teams of two or three users and one professional designer were formed to research specific issues and report back to the main group.

The primary purpose of these meetings and the change in strategy was to get users involved and to help them understand the system.

Summary of Steps. In the preceding example, there were several key steps:

- Management recognized a problem with the conventional approach and sought help.

- Top management was willing to meet with the users and others in a review session. Management was also willing to explain its reasons for undertaking the system.
- At the conceptual walkthrough of the system, there was widespread participation from all areas affected by the system. The presentation clarified that the system was not yet "cast in concrete" and thus encouraged changes.
- The walkthroughs continued, with further attendance and support by top management.
- The initial meetings were followed by action—the formation of design task forces to resolve specific design issues.

Manufacturing System

In this example, the company, a small manufacturer of women's garments, was implementing a computer-based system. At the time this study began, two service bureaus were used: one for payroll processing and the other for accounts receivable. There was a scarcity of information on orders and production planning, however. Existing information processing procedures, particularly in the office area, were insufficient as a result of huge increases in sales volume.

A professional consultant was retained to study the manufacturer's present information processing procedures. This consultant fulfilled the role of a systems analyst. The initial contact with the president of the company provided good management support.

After several months, it became clear that the office manager would be the user in charge of the project. Unfortunately, it was impossible, because of space considerations and training problems, to provide extra help for users. As a result, a long time was needed to develop the system.

Joint Meetings. At the first design meeting, all potential users in the company participated in setting the objectives of the system. These individuals were drawn from production control, scheduling, purchasing, office management, credit, sales, and order processing. More than 10 people were involved in the design process, in addition to the analyst. The first meeting produced general concensus on system objectives, including order processing, raw materials forecasting, and accounts receivable.

Order processing is an extremely important application, both for timely shipments and for scheduling production. Accurate raw materials forecasts are one of the keys to success in this particular business, as it is very expensive to end a sales season with excess materials. While the batch-processing accounts receivable system in use was satisfactory, it was felt that a new system should integrate accounts receivable with order processing and inventory.

After the review meeting, users began to identify the system output, to define report formats, and to develop the needed input. The analyst developed lists of file contents; users determined the field sizes.

Hardware/Software Specifications. The company did not have its own computer at the time and planned to develop the specifications for a system and put them out for bids. Since a batch service bureau operation could be selected, no assumption was made as to the mode of processing. For the most part, a batch-oriented system was designed since it could easily be converted to online input and output, while the reverse was not necessarily true.

As the design proceeded, crucial decisions, particularly in the area of shipping, were discussed in the main group. Alternative scenarios for different decision areas were discussed. Specific decisions of a more parochial nature were discussed in smaller meetings. For example, since accounts receivable was primarily the concern of the office area, production control did not need to spend time discussing detailed accounts receivable questions.

Before submitting a finished document for bidding, two full reviews with the entire design team were held. Again, critical decisions were discussed, and the draft of the system was distributed to the team. After careful consideration, an online minicomputer system developed by a turnkey vendor was selected. Because the system was to be in-house and online, the opportunity existed for reviewing the programming specifications to convert to online input and to eliminate some of the reports with online inquiry. The original consultant who acted as the systems analyst continued during that time to interface the turnkey systems group with the manufacturing company. During this process, the users seemed well informed about the capabilities of the system and its objectives.

Summary of Steps. The example just described illustrates the steps discussed under Adopting User-Oriented Design:

- Although no formal committee existed, a group of key users defined the decision areas to be included in the system.
- The user group itself became the design committee; key users (including the president of the company) were aware of and involved in decisions about the system.
- A user was in charge of the system. Although the analyst was responsible for putting together the documentation on the system, the user was in charge of the detailed decisions reflected in the documentation.
- Management was unable to provide extra resources to aid user involvement, but during the project, a production control supervisor was added to facilitate the development of specifications.
- The analyst acted as a catalyst in the design process; alternatives were explained, and, in general, the users' solutions were accepted. When the users' wishes were very difficult or very expensive to implement, they were very reasonable in making compromises.
- Following the user-oriented design approach, an overview was obtained; then the output was identified, and the input and files were specified. Each of these components was defined in increasing detail through successive iterations.
- Frequent review meetings were the rule. Small groups met to discuss each aspect of the system, while a larger review group met to examine

the entire system. Since individuals served in the review group and a small group, they had good knowledge of one aspect of the system and a working knowledge of the entire system.

- The design was challenged in a manner that was nonthreatening to the user. Questions were asked about whether specific reports or fields were needed. The contractor who was programming the system also challenged the design, asking questions about whether certain information was necessary and whether it was economical to store it.
- The user interface is currently being designed. The online components of the interface will be tested carefully with users. The basic input format was produced by the users, and they will have a strong influence on the screen formats for input and inquiry.
- Conversion is still in the planning stages; however, based upon the knowledge indicated by users so far, all parties are optimistic.
- A post-implementation audit will be conducted after system installation.

CONCLUSION

Better-quality systems should result from user-oriented and user-controlled design because users know their procedures and can suggest ways to improve them. Users become better prepared for conversion and installation and more knowledgeable about the system than when conventional approaches are employed. Finally, users become enthusiastic about the new system—something rarely seen with conventional design techniques. Designing systems according to the approach recommended in this chapter may take longer and cost more, but given the poor record of conventional approaches to design, the increased cost and effort seem well worthwhile.

References

1. Anthony, Robert. *Planning and Control Systems: A Framework for Analysis.* Division of Research, Graduate School of Business Administration. Cambridge MA: Harvard University Press, 1965.
2. Lucas, H.C., Jr. *The Analysis, Design and Implementation of Information Systems,* 2nd ed. New York: McGraw-Hill Book Co., 1981.
3. Lucas, H.C., Jr. *Toward Creative Systems Design.* New York: Columbia University Press, 1974.

Bibliography

"Defining an Information System." *AUERBACH Data Processing Management Series.* Portfolio No. 3-10-01 (1976).

"Performing Systems Analysis." *AUERBACH Data Processing Management Series.* Portfolio No. 3-10-03 (1976).

$\boxed{7}$ Organizational Decision Making and DSS Design

by Stephen P. Taylor

INTRODUCTION

The design of Decision Support Systems (DSSs) is one of the most challenging activities facing DP professionals today. The technological advances of recent years, coupled with the declining cost of DP technology, have permitted increasingly complex problem resolution by automation. Unfortunately, the rapid growth of computer-based information systems has resulted in numerous problems, particularly in systems design. One of the most common difficulties facing DP professionals is the discrepancy between what the user requires and what the DP professional delivers.

The design of DSSs is especially problematic because the demands placed on the system vary significantly from those placed on a simple transaction or accounting system. Decision-making activity within an organization occurs in a largely unstructured environment of constantly shifting goals, priorities, and decision-making styles. Moreover, decision-making activity is not easily analyzed or reduced to a simple equation. Traditional approaches to the design of decision support tools have proved inadequate; new methods and procedures that are based on a more thorough understanding of organizational behavior are needed.

The design of a DSS requires a firm understanding of the decision-making process within an organization. Training of DP and management personnel, however, largely ignores this important topic. DP professionals and business managers thus are often unaware of how the organization actually functions. The designers of management decision systems have traditionally viewed decision making as a rational exercise. The principal decision makers in the organization are seen as logical people who want better information on which to base their decisions. Thus, the systems designers have emphasized improving the predictive qualities of DSS models, providing faster hardware and more efficient software to increase response time, and producing reports faster to improve the immediacy of information. The implicit assumption is that better data and more accurate models result in a better DSS. Such thinking can lead to better predictive tools, yet this alone does not guarantee that the system will function as an important decision-making aid.

Organizational changes resulting from the introduction of a DSS should be anticipated and incorporated into the system design. Such changes should not be dismissed as personnel problems and therefore outside the domain of systems design. The probability of a system's success can be greatly increased if its adverse effects on the organization, the work group, and the individual are considered and minimized during system design.

In addition, the organization's decision-making procedures are a major factor in determining the requirements of a computer-based decision support tool. DSS design will be improved, and the DSS will gain more complete user acceptance if it matches the de facto decision-making process of the organization.

DSS designers must determine which model best fits the decision-making process of a specific organization. A few observations on this problem are made in the following discussion.

DSS Requirements. The organizational context surrounding the system will vary substantially according to the system's function. Systems designed to predict the impact of economic factors on the rate of inflation are likely to have less volatile organizational side effects than do those whose goal is to determine the quantity of resources necessary for welfare programs and the allocation of such resources. The social, economic, and organizational context within which a system will function must be understood by DSS designers. The differences between the public and private sectors as well as those between Fortune 500 corporations and small, concentrated businesses are critical factors in DSS design.

Decision-Making Level. The level of decision making that the DSS will support (i.e., operational, managerial, or strategic) should also be investigated. Low-level decisions require less scrutiny in design than do policy decisions that have significant impact on a large number of interest groups.

Decision-Making Style. Although broad parameters such as those just discussed can be helpful in determining the correct organizational approach to DSS design, the first step is to understand how decisions are made in the organization. Three perspectives on organizational decision making can facilitate this inquiry. Each position highlights certain components of decision making while de-emphasizing others. Together these perspectives provide a much richer understanding of the decision-making process than would be gleaned if each was analyzed separately.

THREE PERSPECTIVES

The first obstacle to understanding organizational decision making is the number of theories on the topic. Academic literature discusses decision-making theories ranging from the normative rational perspective to the descriptive political paradigm put forth by political scientists. Three perspectives on decision making are presented here: the rational, the organizational process, and the political. Each suggests different factors on which decisions are based and thus alters the motives for adopting computer technology as a decision support tool.

The Rational Perspective

This approach has been the classic template for constructing DSSs. Founded on the free-market idea, it defines the organization as a profit-maximizing entity that depends on cost/benefit analysis for every decision. The decision maker chooses the alternative that produces the most utility for the least cost.

In this perspective, decision making can be reduced to an ordered set of steps. The individual is confronted with specific alternative courses of action, each of which is evaluated and assigned an outcome. The decision maker then ranks the consequences to determine the most beneficial outcome. In the case of business decisions, the decision criterion is generally the profit motive.

The Organizational Process Perspective

In response to the rational perspective of organizational decision making, theorists have conceived a model based on the actual behavior of decision makers [1, 2, 3, 4]. This is the organizational process model.

In this perspective, the cost of obtaining all information necessary to make optimal decisions is considered prohibitively high, and such decision making is not considered possible in the real world. Choices are not always clear cut and involve many subjective factors that cannot be stated explicitly in cost/benefit equations. The organization therefore pursues a decision-making strategy designed to produce satisfactory—not optimal—decisions.

Decision-making criteria may depend more on social than on technical factors. Decisions may not be based on their technical value in attaining a goal but on the most acceptable strategy for maintaining the status quo, protecting the interests of the decision maker, and preserving group autonomy and freedom. Such objectives are attained through the development of rules, regulations, and standard operating procedures (SOPs) that reduce the decision-making function to a routine.

C. E. Lindblom suggested that policy formulation is a slow, incremental process [1]. Decision makers move from problem to problem and avoid drastic changes in favor of small, measured steps. The process can be characterized as decision making by successive limited comparisons, where changes are compared in order to arrive at the most appropriate short-term decision.

Herbert Simon's theory of "satisficing"—roughly defined as a combination of satisfying and sufficing—significantly modified the rational model [2]. Simon suggested that the decision maker operates in an environment of "bounded rationality," where the individual's rational decision-making abilities are bounded by a limited ability to perceive, understand, and manipulate

the social world. The decision maker "satisfices" by taking the first acceptable solution found after making only a moderate effort.

The Political Model

The political model sees the organization as a collection of parties acting
independently to further their own goals and enhance their status. The
achievement of individual objectives is put before the pursuit of the rational
goals of the organization (i.e., profit maximization).

The decision-making process is perceived as essentially pluralistic. While
the rational concept may hold for simple heuristic games and the organizational process model for customer accounting and inventory, neither describes
decision making at a strategic or policy level. Many decisions are made in
relation to political constraints, aspirations, and interactions [5].

Viewed from this perspective, decisions result from the interaction of
individuals who focus not on a single strategic issue but on many diverse
problems and who do not act according to a consistent set of strategic objectives but according to conceptions of national, divisional, and personal goals.

Power is the dominant force in the political model. Those who possess the
greatest amount of power ultimately determine the alternatives that will be
viewed as realistic, the consequences that will be seen and ignored, the size of
the stakes, and the structure within which the decision is made.

DSS DESIGN IMPLICATIONS

Computer-based technology can affect the organization's decision-making
process by [6]:
- Altering communication flow and content
- Increasing managerial control
- Shifting power among organizational subunits
- Changing the organizational structure
- Shifting the decision-making function from one management level to
  another
- Psychologically affecting individuals and work groups

Figure 7-1 summarizes the importance given each issue in the rational, organizational process, and political models.

The technical goals of system design—flexibility, reliability, security, and
so on—also greatly depend on the decision-making environment. Each
decision-making model supports a different definition of a "good" decision,
the activity required to make it, and the criteria on which it is based.
Figure 7-2 presents a subset of possible design goals or system characteristics,
together with their relative importance in each decision-making environment.

The Rational Perspective

The rational model presents a normative view of systems design that
stresses the development of purely technical characteristics. The use of mod-

els to represent the external environment and evaluate decision alternatives is of great importance. Such things as speeding the flow of information, obtaining accurate data, and reducing noise in communications channels are examples of important design goals.   *

Decision makers are seen as optimizing solutions within the framework of the organization's goals. The decision-making process is a purely mechanical procedure based on objective and context-independent information; the personnel behind a decision are irrelevant. Conflict among alternatives is not acknowledged. The most utilitarian decision is, by definition, the best and the one taken.

There are some disadvantages to this model. By emphasizing only the technical merits of a system, important design concepts are overlooked. Social interaction and the structure of the organization are not considered. Such factors as power, negotiation, influence, and policy are ignored, as are such design goals as security, programmed decisions, coalition building, and power enhancement.

The value of this model is that it suggests pitfalls to avoid during system design. DP professionals must not construct systems for the purely rational entrepreneur, however. Even if such people did exist, the environment in which they operate must be understood and incorporated into the DSS design.

Effect of DSS on Organizational Decision Making	Rational Perspective	Organizational Process Perspective	Political Perspective
Change structure of organization	Homogeneous decision maker	Coalitions altered	Alters balance of power
Alter flow of information	Unimportant	Vital to subsystems	Political instrument
Affect work groups	Not treated	Subunits affected	Fundamental
Affect individuals	Not emphasized	Not emphasized	Very important
Produce power shift	No concept of power	Alters coalition structure	Fundamental
Alter level of managerial control	Employees as tools	Develop SOPs	Power leverage
Centralize/decentralize decision making	Homogeneous decision maker	SOPs are changed	Shift in power
Alter who makes decisions	Homogeneous decision	Coalitions affected	Political weapon

Figure 7-1. Relative Impact of DSS Technology

Design Goal or System Characteristic	Rational Perspective	Organizational Process Perspective	Political Perspective
Modeling capability	Fundamental	Does not help make decisions	Not essential to good decisions
Program decisions	Not treated	Fundamental	Avoid routinization
Data access (security)	Homogeneous decision maker	Protects coalition structure	Information as weapon
Coalition-building capacity	Coalitions not treated	Fundamental	Fundamental
Flexible data display	Helps evaluate decision alternatives	If SOPs supported	Persuade, influence
Enhance power or status	Not relevant	Avoid conflict	Fundamental
Reliable information	Indispensable	If SOPs supported	If power increases
Dependable	Fundamental	Manual SOPs handle failure	If power increases
Fast response time	Indispensable	Helps information flow	If power increases
Adaptable	Alter models	Organizations change slowly	Organization in constant change

Figure 7-2. DSS Design Implications

The Organizational Process Perspective

The organizational process model suggests that strategic decisions are determined by coalitions, each of which has its own priorities, goals, and focus. Bargaining among these coalitions and factoring large-scale problems into subproblems are the central decision-making activities.

Organizational goals are established and attended to on the basis of slow and incremental change in the membership of dominant coalitions. The introduction of DSS technology can affect the coalition structure suddenly and dramatically and cause unanticipated problems. Changes in the flow of communication through the formal structure of the organization can upset institutionalized procedures and alter the structure and content of work groups. An example is the impact of a Material Requirements Planning (MRP) system on an organization's management, accounting, and production functions. Information gathering, control, and planning are centralized into one subunit. This concentration of activity causes a power shift, alters methods of management control, and necessitates new procedures to make the system function properly.

The system designer must identify the SOPs of those organizational subunits that play an important role in the decision-making process. It is generally difficult to gain user acceptance of an information system that cuts across the organizational structure or intrudes on territorial rights. Nonetheless, an effective set of SOPs often enables the development of support systems that permit new problem-solving procedures to be developed and accepted rapidly.

The technical features of DSS design are given less emphasis in the organizational process model than in the rational model. Such design goals as reliability, dependability, adaptability, and response time are assigned only a moderate degree of importance. In this model, the decision maker chooses the first acceptable solution to the problem at hand, thus eliminating the need for a comprehensive search for all decision alternatives.

Speeding the flow of information through the organization therefore becomes less critical. The decision maker, according to this model, depends less on current and up-to-date information than in the rational model. Acceptable goals are set, and a satisfactory solution to the problem, rather than the optimal solution, is found. The requirement to anticipate all data and decision outcomes is relaxed. Information sufficient for making a decision is satisfactory.

The importance this model attributes to coalitions in decision making suggests that a successful DSS include a means of supporting coalition building in an organization. Although many case studies support the notion of coalition-based decision making, few DSSs possess this design feature.

The principle of coalition building can be applied to computer-aided design tools. Such a system might be enhanced to include a mail system that would permit the ideas, comments, and suggestions of the development team to be circulated among participants. The system can also be used to arrange meetings, disseminate interface design changes, and reduce the overhead of interpersonal communication that accompanies large software projects. These examples illustrate the potential for coalition building inherent in such design tools.

The Political Perspective

The political model of decision making is important for DSS design, especially since it is so seldom considered relevant. Individuals and groups, although committed to a particular goal, will fight hard for their individual point(s) of view.

The political model implies that DSS technology is adopted to the extent that the power, legitimacy, and status of organizational subunits are enhanced. The computer is viewed less as a tool to improve the quality of decisions than as a means of securing the political advantage of one group or interest over another. The goals of coalitions and individuals are seen as the motivating force behind DSS acquisition.

The flow of information and channels of communication are extremely important in the political model. The introduction of a DSS can significantly affect these structures by shifting the function of gathering and analyzing information from one department to another. This shift creates a class structure within the organization, some groups becoming information rich at the expense of other subunits that become information poor. As a result, managers and employees can become anxious and fearful, not knowing how the altered information flow will affect their situations, and they may resist the introduction of a DSS [7].

In the political model, the risk that DSS implementation can result in the inversion of superior-subordinate relationships is acknowledged [8]. A lower-level manager may "program" middle or top management for political reasons if he or she has control of a DSS terminal. That is, the system can be programmed to cover up real organizational problems and uncertainties and thus elicit from upper management the decision desired by the lower-level manager. The computer is viewed as just another weapon in the decision-making arena.

A DSS must be adaptable and respond quickly to the organizational change that frequently occurs in this volatile political arena. The designer must therefore provide a modular design of system components to facilitate organizational needs that constantly change shape.

Systems must be secure and support the formation of coalitions. Data must be secure to prevent others from exploiting it to their own advantage. As in the organizational process model, a coalition-building program is highly desirable. This permits rapid assessment of the relative strengths and weaknesses of a decision under consideration and determination of who must be influenced if the desired outcome is to be obtained. Thus, negotiation is greatly facilitated.

CONCLUSION

DSS design can be significantly aided by an understanding of the decision-making process in an organization. The three perspectives discussed provide contrasting views on the complex activity of organizational decision making and its implication for DSS design.

The systems designer must choose the model that best fits the organization. The best way to synthesize these models into a usable tool is to adopt a diagnostic approach—the image of a doctor making a house call is not inappropriate. In some cases the political or rational dimension may not be relevant; however, designers must determine this rather than assume it.

Work must be done to develop clearer insight into how the organizational decision process affects DSS design. In addition, tools must be developed to help designers diagnose the specific organization's decision-making process. The future success of DSS design depends on how well these tasks are carried out.

References

1. Lindblom, C. E. "The Science of 'Muddling Through." *Public Administration Review,* Vol. 19, No. 2 (1959), 79–88.
2. Simon, H. A. *The New Science of Management Decision.* New York: Harper & Row, 1955.
3. Simon, H. A. "The Corporation: Will It Be Managed by Machines?" *Management and Corporations.* Edited by M. Anshen and G. Bach. New York: McGraw-Hill, 1960.
4. Simon, H. A. *The Shape of Automation for Men and Management.* New York: Harper & Row, 1965.
5. Allison, G. *The Essence of Decision: Explaining the Cuban Missile Crisis.* Boston: Little, Brown, and Co., 1971.
6. Federico, P., Brun, K. E., and McCalla, D. B. *Management Information Systems and Organizational Behavior.* New York: Praeger Publications, 1980.
7. Hoos, I. R. "When Computers Take Over the Office." *Harvard Business Review,* Vol. 38, No. 4 (1960), 102–112.
8. Reynolds, W. H. "The Executive Synecdoche." *MSU Business Topics,* Vol. 17, No. 4 (1969), 21–29.

Bibliography

Kling, R., and Scacchi, W. "Recurrent Dilemmas of Routine Computer Use in Complex Organizations." AFIPS. *Proceedings of the 1979 NCC,* New York, 1979.

Kling, R. "Information Systems in Public Policy-making: Computer Technology and Organizational Arrangements." *Telecommunications Policy,* Vol. 2 (1978), 22–32.

Kling, R. "Social Analysis of Computing: Theoretical Perspectives in Recent Empirical Research," *Computing Surveys,* Vol. 12, No. 1 (March 1980) 61–110.

Lucas, H. C. *Toward Creative Systems Design.* New York: Columbia University Press, 1974.

Lucas, H. C. *Why Systems Fail.* New York: Columbia University Press, 1975.

⑧ Evaluating Software Packages

by Raymond P. Wenig

INTRODUCTION

There are few new applications being designed and developed for computer systems today. Payroll, customer accounts, inventory, and so on cannot be significantly developed further until a major change occurs in the functional structure of organizations. Most current systems work comprises maintenance or replacement projects.

Systems work, therefore, mostly involves reinventing or duplicating. Although customization is needed for many products to interface to an individual organization, base programs in most applications are the same or very similar.

This chapter discusses reusing existing software packages to form all or part of a new application. Such reuse can supply the following benefits:
- Direct cost savings
- Time savings
- Reduced risk
- Better-planned implementations
- Earlier documentation
- Concentration on changes rather than base structure

Reusable software packages can come from several sources, including:
- Unbundled hardware vendors
- User group libraries
- Software vendors
- Other users

Two major obstacles hinder software use. One is locating a software package that seems to perform the desired application and will operate on user equipment. Another is spending the time and energy to thoroughly evaluate package operation and function.

This chapter covers the latter problem, presenting a comprehensive methodology for reviewing and evaluating available software packages. All pertinent aspects of existing software are covered, including user opinions, programming contents, modification requirements, and documentation.

EVALUATING SOFTWARE FOR REUSE

The user should be able to uncover several existing software products that might service a prospective application with minimal effort. The question then involves how to validate performance, content, and usability of such products for specific user applications.

Answering this question requires a thorough and detailed evaluation of the available system. Some of this can be conducted by the prospective end user; however, much of it will require the skills of a software professional. The software evaluation process is iterative; some brief initial tests serve as a basis for identifying software that merits more detailed (and expensive) evaluation. The overall evaluation of software products should concentrate on the following areas:

- Existing uses, users, and performance
- Adaptability to prospective applications
- Structure and content
- Ease of modification

DETERMINING APPLICATIONS REQUIREMENTS

The objective of acquiring and using preexisting software for a potential application is to save time and money, an objective that should be achieved within the framework of servicing end-user needs. Before committing to acquisition and installation, it is necessary to ascertain software performance and usability in the application environment. To determine these factors, prospective users must understand their needs and requirements for the application software as well as hardware limitations and options. Whether or not to use software packages that employ a particular DBMS must also be considered.

Prospective software shoppers should not count on finding a software product that fully satisfies their needs. They should be especially careful not to allow the operations of an existing software package to influence their definition of systems requirements and operations.

Requirements definition must be done by or for the prospective users before starting a search of available software. The following questions serve as a basis for developing applications requirements:

- What are the objectives of the application?
- What transactions must be handled?
- What documents must be produced?
- What files must be maintained?
- What volume of transactions must be handled?
- What unique steps must be taken in transaction processing?
- What controls must be maintained?
- What inquiry needs must be fulfilled?
- What type of environment must the application fit?
- What future enhancements are desirable? What options?

- Who will operate the application?
- What hardware/equipment limits exist?
- What is the user's level of systems expertise?
- What internal system support exists?
- What security requirements and internal standards must be met?
- What type of DBMS, if any, will be used or required?

Answers to these questions should provide a solid basis for reviewing and evaluating available software for potential applications. If answers are unavailable, additional internal research should be conducted before entering the packaged software market.

DOES THE COMPUTER SOFTWARE PACKAGE REALLY EXIST?

Although obvious, there are important initial conditions in evaluating any software package—namely, does it really exist, and is some organization actually using it? Many good software systems ideas have been conceived and promoted without actually ever having been built and tested. Some systems have been designed but never implemented and some designed and built but never used. Still others are heavily promoted while under development.

A key caution for all prospective software buyers: Never buy an unproven software package unless it can be treated as a research and development risk investment. The risks of being an initial user of a new software system, especially one designed and built for another organization, are far greater than those associated with software tailored to specific applications.

Proving that a software product actually exists and is operative is fairly easy. All potential software package suppliers to be reviewed should be required to provide a total or representative list of currently active user organizations, including specific contacts. The user should then invest the time and energy to call and/or visit one or more users of any software product that appears to meet prospective system requirements.

The amount of effort to be spent in determining existence and use of a software product depends on such factors as package cost and source reputation. A few long-distance phone calls to users of $1,000 to $10,000 software products are worth the expense. Systems that cost $25,000 or more probably merit visits to user installations. If a system costs less than $1,000, it is probably more economical to buy the product on a trial basis and test it in the prospective user environment.

When the value of the software product exceeds $25,000 and the package is a complete operational application (as opposed to a utility or a simple application package), the prospective user should be prepared to make significant front-end investments by visiting current users. Software operation and the environment(s) in which it is used should be studied. Some users might find that they easily invest more in product review and evaluation than in procurement.

The search and evaluation process should be treated as a research project and managed accordingly, using appropriate time-and-effort budget control procedures.

CONDUCTING USER VISITS

Visiting user installations provides the opportunity to evaluate software products in real environments and to meet those who regularly experience its performance. Most users are very close to their systems and are directly responsible for processing, personnel, and, sometimes, software support.

Active software users are usually willing to discuss their systems objectively. The visiting team should concentrate their questioning on the following:

- What is the overall level of satisfaction with the system?
- May the team see a demonstration of how transactions are entered and processed by the system?
- May the team review a set of output reports?
- What transaction volume does the system handle (average and peak levels)?
- How long do operators await file responses (average and worst-case situations)?
- How large are the files? What is the growth rate?
- Has data ever been lost? How was the software recovery mode?
- What major and minor software problems have been encountered?
- What changes were installed in the software? What changes are anticipated?
- How is vendor support?
- What are the operators' main complaints about the system?
- What savings were realized, if any? What expenses were incurred beyond initial costs?
- How efficiently does the system use resources?
- What are the resource requirements (e.g., compile and execution times, memory size)?
- Would the user buy the system again? What would they do differently?

Following the visit, a report should be written indicating the visitors' responses to the questions as well as their impressions of the user organization, personnel, environment, and other important systems aspects. Special attention should be given to unique systems features and to operational elements that do not apply to the prospective user's situation. Any hesitation or negative response should be evaluated further and cross-checked with additional user contacts.

PROFESSIONAL EVALUATION

The first step in evaluating an applications software product should be conducted solely by the user representatives. Conducting the user review initially is important, because if a prospective system does not appear to satisfy user requirements, evaluating the product professionally is of minimal value. The risk involved in user review is that the user might accept a system that is technically poor if it appears to operate acceptably. This is a tolerable risk because a poorly built system usually shows many faults at user operating levels. The corollary of this risk—namely, that a satisfactory operational level implies a solid technical product—is not necessarily true. Professional evaluation of user-acceptable applications software is still necessary to advise the user on customization and changes required in a package; different versions of the system may require different support personnel. It should be noted that widely used applications software will contain some dead code and hooks for future enhancements. This shows vendor planning and is advantageous to the user.

Non-user-oriented systems utilities, service routines, operating systems, and other computer support software should also be professionally evaluated. This evaluation should concentrate on content, quality, and flexibility of the software products.

The evaluation should be conducted by individuals with intimate knowledge of computers, programming, and systems operations. A general computer expert is not as well suited to such an evaluation since computer technology, programming limitations, and operation methods all differ significantly. The differences are often subtle, but they are especially important to the user who expects a system to operate in a specific computer environment.

Professional evaluation of a reusable software product should concentrate on:
- Content and quality of computer programs
- Program flows, controls, and systems interactions
- Input, output, and file structures
- Operational tests and safeguards
- Flexibility and expandability
- Documentation
- Evidence of current structured coding techniques and modular design
- Adherence to reasonable standards and practices
- Use of sound design and programming methodology
- Clear identification of previous systems changes and modifications
- Maintainability guidelines

The following sections provide specific tests and details for conducting professional evaluations of computer software.

EVALUATING THE CONTENT AND QUALITY OF COMPUTER SOURCE PROGRAMS

The most significant components of reusable software are the source programs. Any system changes or enhancements will require planning, designing, and installation in these source programs.

A good software professional should be able to review copies of program source listings and to evaluate quickly system content and quality.

Program evaluation consists of a survey and soundness review of certain aspects of software design and construction. It should logically, methodically, and exhaustively review package attributes—both seen and unseen—and concentrate on product operation, maintainability, and extensibility.

The program survey should begin with a quick review of the total package, including:

- Original authors and history
- Major upgrades and changes
- Existing users
- Software maintenance procedures
- Documentation
- Design considerations
- Contacts
- Revisions in process
- Reported errors
- Support procedures

After a first pass through major areas of the software package, the professional review should focus on specific system content, operational logic, and support provisions. The familiarization process should identify:

- System completeness
- Potentially weak areas
- Ease of understanding package constructs
- Consistency

As a final test, an automated source code analyzer (e.g., OPTIMIZER III) can determine the amount of dead code, unused variables, and embedded loops. This will require coordination between the software vendor and the acquirer of the package. If the user environment does not have access to one of these analyzers, a third party may be necessary. A service bureau would be a good resource to tap.

Language

Program source language is key to such factors as ease of modification, efficiency of processing, transportability to other computers, and user independence. The evaluator should check the specific language that was used for the source programs and verify how many of its special features are exercised in the program code.

Strong preference must be given to the use of high-level procedure languages. Ten- to twenty-percent assembly language for application programs is acceptable for reasons of efficiency, but a greater percentage makes it difficult (if not impossible) to modify, enhance, or transfer the system to another computer for a prospective user. Operating systems, utility programs, and special service routines (e.g., data communications protocol handlers) are built for specific computers and must take advantage of machine-level options and efficiencies.

In reviewing the language used in a specific series of programs, the evaluator should become familiar with the vendor's language specifications, especially special features and extensions. The evaluator should then review the program source listings and determine the number of unique special features and extensions used in the system. This information should be available through program documentation or early program comments. In systems not originally designed for reuse by others, these details might not be recorded. In this case, the evaluator must review program source statements to determine the use of special language capabilities.

Some preference should be given source programs that adhere to standard language attributes, since these systems can be more easily transferred to other equipment and users. The greater the use of unique language extensions, the more difficult it becomes to transfer and reuse the programs (except on duplicate vendor computers). A distinctly negative evaluation should be given to programs that use the source language in clever, sophisticated, or machine- or data-flow-dependent ways. Such implementation usually leads to trouble in reuse situations because of the difficulty of modifying the programs without risking the stability of the sophisticated structure. Keeping program code simple, straightforward, and logical is absolutely necessary in reusable software.

The language used in packaged software has several other important implications in the review and acquisition process. Changes in software languages are an example: it is possible to acquire a system whose language is moving toward extinction and diminishing support. Although such a package may operate successfully today, its long-term implications indicate costly maintenance, lack of adequate programming talent, and a distinct generation gap.

The popular high-level languages most likely to survive include:
- BASIC
- COBOL
- FORTRAN
- Pascal
- RPG
- APL
- PL/1

In addition, some new languages are likely to experience long-term survival and growth:
- C
- ADA
- FORTH
- MUMPS

The dying languages include:
- ALGOL
- AUTOCODER
- JOVIAL
- Assembly languages (e.g., Assembler, BAL, GMAP)

In addition to preferring the more popular languages, validating the level and use of language standards in the package is important. Several languages mentioned previously (e.g., COBOL) have multiple standards, which undergo continual change.

Preference must be given to a packaged system using a current-level standardized language. Older standard-language levels become extinct, and many new versions are major, incompatible rewrites of the total language.

Program Comments

Some guidelines and an understanding of the style and approach used in the programs are necessary in source program review. A good programmer can eventually fathom someone else's program but not without wasting much time and money. The best guidelines for understanding and/or modifying existing computer programs are clear, concise program comments. The comments are nonexecuted English definitions of what is transpiring in the program.

A good application program should be filled with clear and concise comment statements. These statements should exactly define the operations occurring in the program and specifically identify the names and uses of major variables. The program evaluator should carefully check the meaning and consistency of program comments.

A suggested approach for checking comments in programs is to randomly select a sample program and obtain a current listing of its source statements. The evaluator should then read this source listing thoroughly. The evaluator should note any unanswered or confusing segments of the program. If necessary, the comments should be reread. When finished, the evaluator should be able to define, with the aid of the comments, the basic meaning, purpose, and operation of the program.

After reading the comments in some programs, the evaluator may be confused. This usually indicates a poorly documented program or inconsistent internal documentation. Such confusion may occur when the program developer has written comments in shorthand or depended on symbols or variable names to denote what is happening. If the reviewer cannot easily understand the program from the comments, it is probable that future programmers who try to change or modify the program will encounter similar difficulties. That it will cost more to modify or enhance such programs is also a fair assumption.

Variable Naming

Another major area of an application source program that should be reviewed is the way in which variables have been named and identified. To change a computer program, it is necessary to identify the variable(s) to be changed and/or to create new variables that adhere to the program flow. When using another organization's program, it is necessary to use the variables and variable naming conventions already established within the system. This means that the clarity and structure of variable names become significant factors in the ability to reuse the software.

On starting the source code review, an evaluator should be able to quickly and clearly identify the type and format of variable names used throughout the program. It always helps if the programmer has provided some embedded comments describing how variables are named, created, and used.

There are three schools of thought on how to name variables in computer programs. One, the "funny school," uses names of friends, enemies, birds, or any other handy, idiosyncratic device for creating and identifying program variables. Since adherents to this school do not produce software that can be readily reused or modified by others, "funny" variable naming should be avoided.

The second school of variable naming creates a meaningful set of 3- to 6-character names that indicate variable meaning and use. An example of such variable names would be the definition of a customer identification code by one of the following: CUST, CUSID, CUSTID, CID. Programs with variables thus named tend to be easily deciphered if the author consistently uses the identification structure. Difficulties can arise, however, if the encoding has been compressed into short character sets.

The third type of variable naming is to name and identify all variables as part of the system structure. Therefore, early in systems documentation, a complete list of the variable names, definitions, and uses within the system is prepared. Some application systems have even included specific program cross-references and areas of use for each variable. This approach indicates that the variable naming was part of the design process and was structured for the flexibility and reuse of the software. Variables named with this level of consistency throughout the system probably provide the reuser with the best form of variable identification.

Program Module Structure

The next factor to consider in evaluating a reusable software package is the overall structure of the program modules. Because of the computer's limited direct-access memory storage capabilities, it is usually necessary to break programs into reasonably sized modules and to call them into memory from a secondary storage device (magnetic tape or disk) and overlay the previously used module(s). Execution control, data values, errors, and so on must therefore be carefully controlled from module to module to ensure that the system properly performs its various functions without losing or erroneously changing the proper results.

One of the first structural areas to check is the number of program modules in the system being reviewed and their relative sizes. If the system only contains a few relatively large modules of 10,000 or more statements per module, it will require significant dedicated computer memory (300K to 400K words or more) for one program alone. Such a structure, usually performance efficient, requires more dedicated resources while in operation. The more desirable structure (especially for interactive, multiple applications systems) has smaller but more modules with logical functional segmentation. As memory costs continue to decrease and speed increases, however, this becomes less of a cost factor.

Preference should be given to a computer system that uses a modular program structure, with each program limited to from 100 to 500 source language statements. Although such systems are usually better constructed, require less computer dedication, and perform interactively, they are somewhat less performance efficient than large programs. Another advantage is that small modules are easier to change and/or replace when modifying or enhancing a system. As programmer costs continue to rise, this factor becomes more significant. Documentation is also usually better because it is easier to produce for small module entities.

In reviewing program modules, the following factors should be considered:

- The relative size of modules should be consistent. The evaluator should check specific sizes from which to build a distribution histogram. The better systems have more modules (+80 percent) within a narrow size range, such as 300 (±50) source statements.
- Common data areas should be well defined and standardized across individual modules. They, too, should be of consistent size, appear in standard positions in the program, and contain logical variables.
- Logical file structures should be used for defining, building, and storing data. They should be defined so that variables, sizes, types, and contents can be quickly determined. The file definitions are most flexible when done on early systems modules and then reused via a shorthand reference.
- Consistent error handling methodologies should be used in the program modules. This would include clear identification of errors and error messages as well as clear audit trails on error disposal or repair.
- Transaction audit trails should be maintained by all modules whenever a data file is changed. The audit files should be available for restarting the system, developing history profiles, and testing and certifying system performance.

Interface Linkages

Once the content of the program modules has been checked, the evaluator should study the connection methodologies between modules. Many computer systems (except small utility systems) are built on a module overlay concept: either a base master control program is used to call in appropriate processing modules, or each module has built-in logic capabilities to select the next performing module.

The evaluator should check that the calling sequence for program modules is clear and standardized. All pass-along variables should be clearly identified and stored in a common data area or explicitly named in a transfer section or calling sequence. A few randomly selected modules should be reviewed to validate linkage consistency. Special attention should be given to value-passing through the use of absolute address locations, implied-value strings, trailer values attached to common data space, and other nonstandard programming practices. Passage linkages are an indication of tricky programming, which means that the programs may explode when someone tries to modify and/or enhance them.

Good interfaces are major factors in judging the adequacy of system design and program construction logic. Proper linkage construction involves:

- Clear module naming conventions
- Well-defined parameter passing processes
- Good documentation on all interlink relationships
- Definition of linkage entrances and exits
- Documentation of purpose and operation
- Cross-reference listing of all to/from linkages to other modules

If the linkage-interfacing process is complicated or unclear, the software packages value should be downgraded. Later extensions to the system will have to be interlinked to the existing process. Any ambiguity or excess complexity in linkages will hamper package extensibility and greatly increase enhancement costs.

Restart Provisions

If software fails, operators must be able to restart programs quickly and correctly and continue their processing without loss of data or damage to master files. To accomplish this, the software system must have adequate built-in restart capabilities and automatic storage of necessary recovery data.

The professional evaluator should carefully inspect the software modules and documentation to determine how, when, and where any restart provisions are active in the system. Once located, their operation and expected performance should be thoroughly evaluated in an actual processing environment.

The best type of restart provision is a continuous data audit trail file. This approach maintains a complete, continuous log of significant transactions against major files, plus regular recording of significant program values at main overlay points in the system (e.g., after file updates, long processing, or computational modules).

A check of the restart-data file layouts should confirm that necessary data is present to support a systems restart. Consistency of restart-data generation throughout the system should be checked.

If time permits, an actual test of the restart process is worthwhile, especially when the evaluator has some doubts about the apparent programming of the restart process.

File Structures

File structures are significant elements in the overall software system architecture. Most transaction processing application systems actively build, access, and depend on their data files as a major reference and operational link in their processing. Many program modules might be working in the total system, and the file structures represent a major element that knits together the system.

It is possible to change systems programs without affecting the file structures; however, it is seldom possible to change file structures without causing several program module changes.

The file structures are crucial common keys that usually support many program modules within a system. The evaluator should ascertain that detailed file layouts exist for each file, with a clear explanation of the contents of each field. The next step is to locate the actual file layout definitions in the software listings and validate that the documentation and the programs are consistent. A random check of two or three major files should suffice.

A DBMS-based environment demands slightly different considerations. This is especially true if the software package uses a different DBMS or none at all. The conversion effort must then be factored into the decision process.

Each major file should be closely studied to determine its structure, organization, keys, expansion capabilities, audit-trail generation, health tests, and overall flexibility. The following are some of the factors to be reviewed:

- Keys—position, size, unused fields, check digits, access authorization, subkeys, links to related keys.
- Audit trails—date and type of last record change, old/new record logging, use counts.
- Health tests—record bit counts, use bit tests and comparisons, bit discrepancy handling.
- Expansion capabilities—expected ease of changing or adding fields to the records, including unused embedded or end-of-record fields and the ease of changing file-definition programming.
- Linkage structure—resolution of duplicate record keys; connection to subordinate records; ability to locate partial, embedded, or multiple keys.
- Organization of the record—physical chronology of data fields in order of use or importance.
- Handling of unfilled fields—used for storage of unfilled data fields in a record (e.g., space filled, zero filled, special-character filled).
- Variable packing—used on numeric values in the file, specifically whether they are in binary, decimal, or packed decimal format.
- Character representation—used for storing alphanumeric data in records to determine whether it is EBCDIC; ASCII; full 8-, 7-, or 6-bit code or whether any parity bits have been removed to compress space.
- Equipment dependence—computer file devices (magnetic disk, in particular) often have fixed vendor-oriented limits on record sizes, key structures, and other elements. A check of the files and the programs should be made to test the level of equipment dependence (or preferably independence).
- File types—keyed, indexed, variable, fixed, blocked, spanned; determine which is used and how equipment dependent it is.

File structures should have consistent physical structure and programming-level definitions. Preference should be given to easily understood, logical file structures, rather than to complex ones.

Software packages using separable file management routines or a data base approach are best: they allow easier modification of file structures and content without requiring modification of each routine. Although few standards exist for file or data base management systems—many of which are unique proprietary packages—they can enhance maintainability of the software product and increase adaptability to user requirements.

Fail-Safe Operation

Once installed and operating, an interactive computer system must be able to continue operating no matter what. This is critical because most computer applications will be operated by non-DP personnel. The system will usually not have a protective layer of specialized operators. The applications software must tolerate all types of difficulties and recover to a position from which the operator can restart the process.

Power failure, data file destruction, illogical and improper data entry, or data file or program intrusion by an unauthorized user are examples of computer system failures. Proper computer programming controls must provide support for these potential problems. Some failures receive hardware support, as in the case of power-failure detection. Even here, however, the programs must be able to recognize power failures to prevent data loss and to provide instructions for recovery and continuation once full power is returned.

Fail-safe operation should be treated as a critical software attribute. The professional computer evaluator should verify that there are connections at all possible fault points to test for proper system operation. Test results or default options should always transfer control to a restart point, saving all necessary data values. Erroneous or illogical data entry could cause a transfer by a rejection routine that removes the erroneous data from the system, logs the errors, and then transfers control to a clean data reentry point.

The evaluator should also check that the system cannot lose control between program modules or be forced to subvert the logical order of modules. Preferred systems are those with built-in checks and balances between modules or control counters that use values generated by prior modules as keys for next-step processing.

Hands-on access to an operating version of the system may be expeditious for the professional evaluator. This can be done at an existing user installation or at the software vendor site. Such operational evaluation should consist of running the system and attempting to jam, stop, hinder, disrupt, or destroy system operation.

The potential user should not be too discouraged if the software package is easily destroyed in actual operating conditions. Few current computer application systems would receive very high ratings in this category. This is a significant and important area in evaluating an application, but it is one the evaluator will find a pronounced weakness. Although a system without high ratings in this area does not warrant automatic rejection, it can cause concern since it affects the operational use and long-term stability of the system. For systems without built-in fail-safe capabilities, the user should evaluate the costs required for their addition.

EXTERNAL ENVIRONMENT: USER INTERACTION WITH THE SYSTEM

The end users of an interactive software package must manage and control the system's overall flow and performance. The professional evaluator should review the software to verify that the system can properly interact with its users. Key areas are discussed in the following sections.

Systems Start-up. It is important to set date, time, transaction control numbers, security keys, restart controls, and so on.

Program Selection. This includes the process of reviewing available program action choices (e.g., through screen menus) and user input of chosen actions.

Error Messages. The format and display of error messages (e.g., the use of reserved error area and blinking screen messages) and the clarity of typical messages (abbreviated codes versus lucid definitions of the errors) should be checked.

Error Corrections. These are procedures for inserting proper values into the system in place of designated errors, including instructions to users concerning error correction options, handling of error data (e.g., outright rejection or holding in a suspense area awaiting correction), and software design philosophy regarding error audit trails and ultimate disposition of errors.

Operator/User Instructions. Comments to help the operator correctly and efficiently address and use the system should be clear and concise. The length and detail of the comments, the availability of detailed backup instructions, listing of alternative choices, and the use of shorthand or omitted messages for experienced operators are all important aspects of this task.

Operator Action Commands. These are the level-of-command user entries that cause the system to perform a chosen action. They may vary from single-key entry commands to complicated sequences of multiple-character answers. Single keystroke commands offer simplicity and speed; complicated sequences achieve better security control.

Restart Procedures. These are the actions necessary to reestablish proper system operations if the normal process is interrupted. Included are halt indications and instructions given to the operator to restore the system to its correct working position. Restart procedures should include identification of last-accepted data values and the required place to restart the process (e.g., by indicating any requirements to reenter lost data).

Descriptive Data Overrides/Exception Entries. These involve the operator's ability to override selected data fields in an input format and to enter exception input into data files. The evaluator should review the override process and ascertain whether it adheres to logical rules and controls. The normal approach for descriptive data overrides is to allow the user to make a specific entry in every field in a format by stopping the terminal cursor at the start of each field. If a user chooses to make no entry, he or she presses the skip or line-feed key. The system then places a default value into the skipped field. The alternate method is to place the default value in the data field first. The user is then allowed to place a substitute value by tabbing the cursor to the field and entering the override. Either method is acceptable. The evaluator should determine that a sound audit trail is kept on such changes and that controls for review and/or authorization of the exceptions are properly implemented in the software (e.g., supervisory terminal reviews, data override reports, detailed transaction audit trails).

File Updating. The steps and control procedures involved in proper application of revised and new data to master and work files are critical areas of any data system. It is important that the software package perform its file updates:

- In a user-transparent mode
- While adhering to accepted audit standards
- While providing for safety points in the event of a hardware failure
- While checking all data processed for consistency and errors
- While validating the acceptability of update results
- While maintaining an indicator of the status of update progress

The conclusion of any file update process should verify the number of records processed against an initial count and provide a set of file content values. Hash-total checking on updates is considered a plus factor.

DBMS Considerations. The current operations environment significantly affects the requirement for a data-base-oriented software package. If no DBMS is currently in use and one is desired, selection is the primary concern. Then the market can be surveyed for an application package that uses that particular DBMS. It may be difficult or impossible to find such a package, in which case conversion may be necessary. The evaluation process must then include a step to estimate the effort needed to modify the new software.

Report Generation. The process of generating systems reports involves controlling when reports are generated (i.e., whether a logical process, such as file updating, must be complete before reports can be called out). It also involves obtaining the user input necessary to define the report details and outlining the operational steps needed to produce the report.

The user inputs (e.g., report name, data, and control values) should be simple and straightforward. Most of these values should be automatically created by the report generator, with allowance for user overrides. In fact, many reports may be automatic final steps in a systems process, without any requirement that a user specifically request their generation.

Producing hard-copy reports necessitates evaluation of the software's printing capabilities, including report alignment output and operational steps to be taken to ensure proper positioning of the data on the paper. Good software features in this area will include the optional ability to selectively reprint individual reports based on types, number, or other criteria, without reexecuting the system (i.e., a spool routine).

Inquiry Procedures. Most interactive systems need to inquire into data files to ascertain the current value/status of various elements. The inquiry process involves user identification of needed data and system retrieval and display of the proper results.

The evaluator should review in detail the inquiry-input process—including menu selection steps used to identify the type and characteristics of the inquiry, input screen layout, guidance messages, partial or subordinate identifiers, and other retrieval definition factors. The disposition of the answers should then be reviewed (e.g., whether they appear on the CRT or printer, user options on details to be displayed, and whether hard copy can be requested).

The evaluator should ensure that the inquiry software does not change any file values and that the inquiry process can be ended and system control returned either to a neutral position (i.e., program-selector menu) or back to the interrupted processing. The evaluator should also determine whether the requests are logged to the transaction audit trail.

Exit Procedures. These procedures are used in performing a normal closeout or an emergency exit procedure of an action process. In normal cases, the software should allow an easy exit from the process when a user indicates the end of data has been reached or explicitly enters an "end" action command at a proper sequence point. The emergency exit procedure must allow proper logging of in-process data, generate restart positions, properly suspend actions, and transfer control to an action-selection module. A review should be made of the software's method of passing system control to the next-logical modules.

Shutdown/End-of-Day Procedures. These are operator instructions that indicate the logical end of a unit of processing (e.g., shift, day, week, or a particular function of the system). These procedures should be automatically enforced before a next-logical-unit process can be instituted (e.g., a date change for the next day's start-up). The shut down process should be checked for proper generation of audit logs and the necessary procedures for required action steps.

Interactive Flow Control. The processing of most applications software packages occurs on a partially or fully interactive basis on most state-of-the-art software packages. The interactive processes should be carefully defined and easy to follow. The control over sequence and actions could be automatic, through user menus, by terminal function keys, or through user commands. The major considerations for interactive flow controls include user initiation, activity progress indicators, error handling, interrupts, and restart provisions.

User Documentation

User documentation is a deliverable that helps take the place of resident software experts. It therefore must be complete, easy to use, well indexed, and accurate. It should also be well written, professionally reproduced, and easily updated. User documentation should include the following elements:
- An overview of the software product, its structure, and application objectives
- An explanation of input procedures and data items, including edits, tests, errors, and control steps
- An explanation of output reports and data items, including a definition of their source and/or deviation
- Step-by-step operating instructions, including start-up, recovery, shutdown, program calling, and audit tests
- A failure analysis matrix with recommended action steps for restoring proper processing
- Maintenance request procedures and service response events

Output Reports

Most software produces a series of output reports as a major product for users. These reports are also one of the major design features of the software product. The content, layout, flexibility, and usability of the output reports represent significant characteristics of the package.

The professional evaluator should review the reports in terms of overall structure, logic of layouts, headings, totals, controls, and other factors. He or she should also check the report writing programs to determine the degree of ease or difficulty in implementing changes to the reports.

In addition, the evaluator should review the options and external choices that are built into the software. These would include:
- Report headings
- Field size limits or flexibility
- Control numbers
- Distribution identification
- Sort selections
- Report suppression
- Choice of levels of details
- Types of error messages
- Selection of total breaks
- Content options

The more of these parameters that are built into the software, the easier it may be to tailor it to user needs without expensive program modification. Such flexibilities may increase the difficulty of installing report changes at the programming level, however. A careful analysis of the reports' end-user suitability and of the types of expected changes will aid the evaluator in judging the flexibility of the output report programs for the prospective buyer.

More generalized software may feature a built-in report generator. This capability can be a great asset in helping end users obtain tailored output at minimal programming cost. If this capability is present, the evaluator should review its structure, use, and flexibility. The following areas require special attention:

- Field selection
- Position indicators
- Total generation
- Line counts and page breaks
- Field/column headings
- Data controls
- Ease of use

Some report generators are little more than facilities to call an open program subroutine. Someone must then create the programming steps—data selection, computation, totaling, positioning, and other detail-level steps. If a report generator is included in the software, the evaluator should walk through an actual sequence of report building and generating instructions to ascertain the capabilities and flexibilities of this feature.

Output Forms

The generation of transaction reports from packaged software may require the use of special preprinted forms. These could include:

- Checks
- Picking lists
- Statements
- Purchase orders

The acceptability of such forms to end users should be evaluated. Some package vendors are the primary source of the forms and use them to establish a lock-in relationship. Conversely, some define the format but do not supply a sample. The users must design their own and contract for forms supply.

A check should be made on form flexibility and the ease with which such variables as fields, location, and content can be changed to ensure usable package outputs. The use of multipurpose forms or user-defined formats should be considered a package asset.

Computer Output Microforms (COM)

Micrographics are increasingly used as an output medium. Many systems now have specific files designated for COM, while others contain hooks for COM software routines. The evaluator must know whether the present environment is roll film or fiche, online or offline, and must be able to define the minimum requirements for the new software package.

Systems Documentation

Most software buyers are concerned primarily with operating and user documentation. Systems documentation, which contains significant information about the logic, structure, and flexibility of the system, is often given only a cursory review. It may be weak, incomplete, and inconsistent, and the weaknesses may not be apparent until the system fails or a user wants to install a major change in the software.

The professional evaluator should critically review systems documentation for the following:
- Systems logic flow diagram
- Narrative overview of the system
- Flow diagram and logic narrative for each system's module
- Readable record layouts with detailed data element definitions, including sizes, edits, and data sources and uses
- Input and output record layouts with definition of elements
- Definition of any program-level options
- Definition of all variables
- Explanation of any open or reserved variables or code sections
- Definition of all audit tests and edits in the system

Systems documentation often consists of source program listings with some program comments embedded in the code. If the comments are adequate, the listings can be considered minimally acceptable systems documentation. A careful check should be made to ascertain that program comments are consistent from module to module and are up to date. At the very least, the embedded narrative should:
- Define all variables
- Identify major decisions and functions
- Outline all entries, exits, and error conditions

Audit Provisions

Audit provisions are critical to checking and balancing data files, testing system operations, or validating financial accounts. If audit mechanisms are inadequate or inoperable, it is difficult to acquire or build good test programs to perform such functions.

The professional evaluator should review the software for its ability to support a reasonable set of audit requests as a by-product of the normal processing steps or as a special processing procedure. Desirable audit procedures include:
- Displaying the value of intermediate variables
- Tracing a transaction through all processing steps, providing step identification and value output
- Selecting and printing specific records from data files
- Extracting log entries of a transaction and tracing its disposition through the system

- Producing a balanced output of all data items in and out of all system modules, by number and amount

The software developers should have included most of these audit capabilities to aid in testing the product. These capabilities should have been left in the system for use in future enhancements validation and for data auditing. If these capabilities are missing from the system, the evaluator should estimate the time and cost to build them. As an alternative, the reviewer might check the feasibility of using a generalized data auditing package on the computer applications software and associated data files.

Competitive Processing

Application systems are real-time, user-demand-responsive systems that support the processing of various concurrent applications based on current user needs or demands. The professional evaluator should test each system's ability to perform under heavy data and user loads for varying mixtures of applications.

Several problems usually become apparent during the competitive processing review, such as:
- Slowdown in user-terminal response time
- Reduced output speeds
- Lengthy queues of requests for file information
- Excessive operating system overhead times

The competitive processing review, therefore, should evaluate the following aspects of the software:
- The maximum time to access the same record from a file if all user terminals request access simultaneously
- The maximum number of concurrent resident applications that can be active under the software
- The estimated maximum rate at which transactions can be entered into the terminal, including editing and error correction times

Evaluation of the final area should concentrate on predicting the operation of the software when maximum loading conditions exist. Such conditions may not occur often; however, when they do, it is usually during a period when users are intolerant of system slowdown or failure to carry the load. The selling vendor should respond to these concerns in writing, and an attempt must be made to tie them to the contractual performance specifications of the software.

Customization

Most software packages require customization to tailor them to user needs. Such changes require careful definition before a commitment is made, since the costs of installing and validating the necessary changes may exceed the cost of the package. In addition, it can be difficult to install the desired changes in the prospective software.

The evaluator should work with end users to review the appropriateness of the various elements of the software product. Together they should define a list of necessary characteristics and a list of options for the software. All user-oriented elements in the product (e.g., input formats, reports, displays, and processing rules) should be reviewed. For each element, the users should indicate the mandatory and desirable changes.

The evaluator should then meet with the developer to determine the necessary effort and anticipated difficulties in making the required changes. The output of this effort should be a work-task test with some resource estimates, including associated time-and-cost values. The developer is often the best source of input regarding changes. The developer has unique knowledge of the product and should be able to produce the most changes within the least time-and-cost framework. At times, the developer can be convinced to provide some customization effort in the quoted product price.

If the software developer is not available to perform any product change work, the evaluator should define the changes in detail and the expected costs to install them. If the evaluator is not qualified to make such estimates, local software firms should be issued an RFP to install the changes.

The overall review of any software product should take place on a complete-cost basis to show the total time and cost to produce an acceptable working product that the purchasing organization can use. Often, it is easier to make major modifications to a less-than-satisfactory product than to make small changes to a more complex system that satisfies all user needs. Program structures, documentation, flexibility, files, report generators, edits, and other building-block modules play a significant role in system changes.

Vendor Support

Any applications software package requires a certain degree of vendor support. Basic support should be provided with the package; additional support should be available on an as-needed basis at an established price.

The areas of vendor support that should be evaluated are:
- Customization of input, output, and options on the software
- Operational installation and setup of the system on the buyer's hardware
- Training of operators and system users
- Audit of system operations after some actual use
- A specified number of days or man-hours to support the package
- Availability of telephone consultation to help in using and understanding the package
- Availability of a retainer type of priority maintenance service to ensure prompt attention and to help keep the system running

The systems software purchaser may not need all of these items. Many software products can be self-installed and easily understood, provided adequate documentation exists. When the selection is finally made, the buyer should be sure that the required and agreed-upon vendor support items are clearly spelled out in the contract. If these areas are not included with the package, or if they are a separate cost item, the prospective buyer should estimate the costs of acquiring this support before comparing the product against one with a heavy amount of built-in support. If the developing vendor does not offer support services, then an estimate of the costs of the internal or external effort needed to provide the required support is necessary.

PACKAGE SELECTION AND ACQUISITION

After a complete analysis, the prospective buyer must select a particular package. Different situations and uses dictate the proper weights for the various factors. The use of numeric rating schemes and plus/minus ($+/-$) lists can be helpful in sorting the competing packages. The buyer must balance the specific factors covered in the previous sections and must judge each vendor's capabilities.

Once the decision is made, a complex and often frustrating postselection phase occurs—contract negotiations. For simple systems, a contract is usually a brief licensing agreement and a purchase order. For more significant software (more than $5,000), the agreement should take the form of a clearly written legal document that spells out the responsibilities and obligations of all parties. Some vendors employ a standard contract. Most contracts, however, are one sided or incomplete. The buyer should ensure that a legal counselor reviews the contract and makes appropriate modifications and/or additions.

The best way to commit to an outside software product is to treat it as a new product investment project. Although the basic package exists, it is not successfully implemented until it begins to service the user organization. Bringing the software to this level can represent a major effort, often many times more costly (in user time and energy) than the review and selection process.

Defining System Changes. As mentioned previously, very few reusable applications packages will fit a new buyer's needs without a significant number of changes. Quite often the cost of the changes exceeds the cost of the software package, and this should not be considered abnormal.

Data Conversion Requirements. The specific software product selected will define a significant amount of the data base and file elements for the final system. Users are cautioned not to attempt too many changes in a software system data structure.

The new system's data structure will become the repository of user data, thus necessitating data conversion from its current internal form and format (e.g., file folders, ledger cards, service bureau files, single records, other computer systems) to the new software structure. This conversion process requires:
- Access to current data values
- Conversion to the new format

- Input into the new data storage media
- Validation of content
- Parallel updating of values
- Cutover to new system operations

The vendor should know how to convert successfully and may be able to supply some level of direct support and help.

The data conversion process should be treated as a special user-managed project that can parallel system modification and installation. The conversion project is probably the most significant commitment the user must make to a reusable software project because it involves a commitment of the organization's human resources to learn a new system and to translate familiar data into an unfamiliar form.

Product Acceptance. As soon as the software product is selected, the organization should define the tests and conditions that will determine the successful operation and acceptability of the finished software product.

Acceptance tests, which should be treated as a multifaceted user project, can aid in learning, certification, vendor evaluation, and future enhancements identification. Acceptance tests should be designed to validate the total system, the weak areas, and any changes that have been implemented in the system.

A major part of the acceptance project is generation of a comprehensive set of test data. The vendor often supplies such data for acceptance testing. Users should review and enhance it or supply their own. This final test data should contractually form the basis for acceptance. The test data should be designed to test all facets of the software, including:

- All transactions
- Errors
- Expected error combinations
- All major reports
- End-of-period processing
- File maintenance options
- Audit tests

Test data should be prepared as a permanent data product used for initial system validation and subsequent revalidation whenever a change is made. For viability, the user organization should also maintain a full set of accurate answers for all test data elements.

The next step in the acceptance process is to develop a detailed validation and testing plan. This plan will become the step-by-step processing and evaluation guide for the acceptance efforts. It should include:

- File-building procedures
- Test data execution steps
- Auditing/results comparison efforts
- Volume testing procedures

The product acceptance project is a demanding and critical effort for a new or modified software system and is the organization's last checkpoint before full implementation of the product. The more thorough the project, the greater the chance for a successful system. The acceptance team should negotiate with the developers to solve all critical problems prior to implementation and to generate a reasonable schedule for correcting any noncritical problems.

Product Documentation. When the buyer selects a particular software product, he or she commits both to the system and its documentation. The organization should review the available documentation and develop documentation acceptable to the users.

Even if the available documentation (especially user-oriented instruction) is adequate, the rewriting of all or part of it for consistency with the user organization should be considered. The process of documentation enhancement provides:

- A review of the system from a user's viewpoint
- "Apparent" tailoring of the product to fit the organization's needs
- Expanded coverage of important system aspects

Additional summary documentation can be helpful in reducing organizational resistance and in establishing a positive attitude.

The documentation enhancement project can occur in parallel with the software modification project. It is a user-run project that encourages participation in the total system and makes knowledgeable users available to cooperate and communicate with the modifications development team.

Implementation and Stabilization. The final commitment to a new software system is the installation and use of the product. The installation process involves the execution of the training program, distribution of documentation, and the collection and processing of "live" data. It also involves:

- Converting existing data files
- Responding to faults, problems, and complaints
- Controlling user uncertainty and fear(s)
- Working with the developer to identify and correct system problems

The implementation process also involves stabilizing the operation of the system. Improving the system's performance is nearly always accompanied by unexpected problems and hurried responses and repairs. The process must be user controlled and requires a dynamic form of real-time project management.

As implementation begins to stabilize, the user team should determine the learning curve for the system and plan the steps to improve user and system performance. The team should also plan the first full system audit and evaluation and determine that the product is performing satisfactorily.

CONCLUSION

Successful implementation of packaged software requires:

- Good-quality software products
- Willingness to compromise on details

- A responsive product vendor
- Time spent learning and practicing the use of the system
- Clear understanding of needs and expectations
- Properly implemented modifications
- Adequate long-term support
- A comprehensive test plan

The reuse of available software can save both time and money, resulting in a better product. The achievement of these benefits requires a great deal of work and investment to ascertain that the product can adequately serve user requirements. The detailed list of checks provided herein should be followed to ensure the promised results from existing software packages.

⑨ Organizing for Project Management

by Leslie H. Green

INTRODUCTION

Project management generally refers to a management process that is designed to deal with a specific problem or achieve an explicit objective. In its simplest form, project management can be defined as assigning personnel within a functional organization to a temporary task force for the completion of a specified task. At a more complex level, project management refers to a highly fluid organization with little hierarchical structure and within which people are rotated in and out of project assignments as required.

Three ingredients are necessary to enable effective project management:
- A project organization that possesses the skills necessary for project completion
- A manager or management function that can integrate the skills necessary to accomplish the project objective
- A development methodology for implementing the project objective

Systems development methodology is the subject of this service as a whole. The other essential elements of project management are discussed in this chapter.

APPROACHES TO PROJECT MANAGEMENT

To accomplish effective project management, one of four basic organizational forms can be employed.

The Functional Approach. The first, and perhaps most common, method is to employ the existing hierarchical organization. That is to say, at some level within the organization sufficiently high to direct all related aspects of the task, a project can be formulated. In a formal or informal fashion, the various reporting functional units are apprised of the project and their required contribution to its completion. Throughout the life cycle of the project, periodic reviews are conducted by the involved functional managers. These reviews are formally or informally conducted until such time that the various subtasks are completed and integrated into the end product. This form of project organization is conducive to relatively small or uncomplicated projects.

Project Teams. An alternate project form is one in which a project manager is appointed, and all resources necessary for completion of the project are directly assigned to him (see Figure 9-1). Within this structure, the project team or task force, under the direction of the project manager, is solely responsible for project completion. Inasmuch as resources from the related functional areas are assigned to the project manager, there is little or no need for external interfaces with the various functional units. The significant difference between the project team approach and the functional approach is that the project manager is able to direct the efforts of planning and implementation of projects without crossing organizational boundaries.

Matrix Form. Another variation on the manner in which projects can be managed is the use of a matrix organization. Within this structure, a project manager is appointed and made responsible for the project; however, resources required for the project are retained by the functional units (see Figure 9-2). This form of project organization is frequently employed in large projects where diverse and sophisticated skills are required and where those skills can best be managed by functional managers more adept and skilled in the

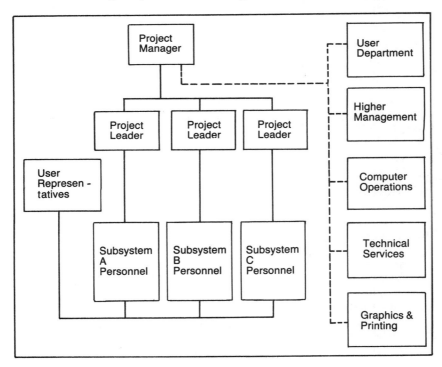

Figure 9-1. Organization of a Project Team

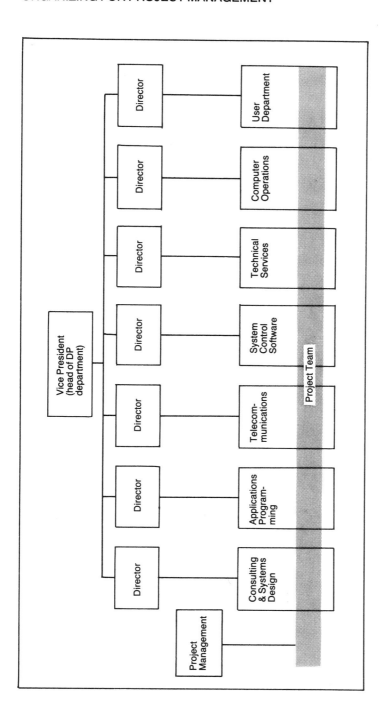

Figure 9-2. Matrix Project Management

various disciplines. In this project structure, a number of advantages are gained in terms of bringing the best efforts and resources to bear on the needs of the project. The appointment of a full-time manager responsible for the management of the project provides the level of attention required for project direction and control. Likewise, the assignment of personnel to the project within the functional areas provides the quantity and quality of required personnel resources, with a minimum of disruption within the functional areas.

Influence Form. A fourth form of project organization is one in which a project manager is responsible for project completion without a dedicated project team or formal interface within a matrix organization. In this form, the project manager acts as a responsible agent for the project. The project manager influences task accomplishment and integration through communication with and through the various involved functional areas. The essential difference between this form of project implementation and the functional approach is that the appointed project manager's sole responsibility is to the project; he is not, therefore, concerned or otherwise deterred by additional managerial duties or responsibilities. This structure can more readily be recognized by the use of such phrases as *special assignment* or *ad hoc tasks*.

PROJECT IMPLEMENTATION

The successful implementation of projects requires that a disciplined approach be adopted. This approach varies, based on the approach to project management implemented in the organization.

Functional Project Management

Projects that are accommodated within the functional organization are most familiar to many businesses. If the requirement for a new product or service is recognized, the chief executive officer (CEO), in discussion with one or more department heads, would acknowledge the requirement as well as commission a project for its development. In this organizational project form, the project is run or managed by the CEO or a designated line manager. Project responsibilities are typically in excess of ongoing line responsibilities and, as such, receive less attention than problems encountered within the spectrum of line functions.

Tasks required to complete the project tend to be generally defined by the manager assigned responsibility, with detailed definition left to the various involved departments. Task definition and implementation are carried out within the functional areas, with little formal regard to project integration except in those areas where coordination with other functional areas is recognized as a prerequisite for task implementation.

The management of the project is likewise decentralized, and few formal control mechanisms are apparent at the project level. Control of tasks within the functional areas is accomplished by the functional manager or managers involved, with little consistency among functional areas.

Project status tends to be reported on an informal basis, and decisions that affect project form or content are made by those areas most affected. Decisions that affect the total project are generally resolved at a higher level, with the manager assigned total responsibility.

Finally, project planning tends to be done on an impromptu basis as problems arise. Overall project plans are infrequently drafted and updated. Little formal contingency planning is evident, except as might be noted within the functional areas as detailed tasks are defined.

The functional project form of organization works well with relatively uncomplicated projects in which project tasks can be easily accommodated by the various participants, and the requirement for detailed planning and integration is not paramount to project success.

The disadvantage of this form of project implementation is that it does not accommodate all projects. In particular, it does not satisfy projects that are technically complex or that require long time frames for implementation. The reasons for this are many; however, the most significant one is that the authority and responsibility for project completion are either too diffused among the participants, or, if focused or centralized, they are vested in a manager with responsibilities in addition to the project [1].

The management of large, complex projects requires an enormous amount of time and attention. Decisions are frequently required and must be made with a minimum of input from project members. Planning, coordination, integration, and control of project elements can overwhelm a functional manager with additional duties, with the result that neither the project nor the line responsibilities are effectively managed.

Influence Project Management

A variation of the functional project form, influence project form [2], attempts to solve the basic problem of the functional form by assigning a project manager to the project effort. Typically, the project manager is staff to the responsible manager and has no direct responsibility for the various tasks being performed within the functional units. In this case, the project manager is more of a monitor or expediter than a manager and integrator of project activities. The responsibilities of the project manager are to call meetings for discussion of status, to act as arbitrator to resolve project conflicts or problems, and to influence the development of the product or service as he deems necessary or in accordance with instructions from the manager ultimately responsible for the project.

Depending on the size and complexity of the project and the political and organizational environment within the functional organization, this form of project management can work well. It solves the problem inherent in the functional form in that someone is committed full time to monitor the project status and to ensure, to the degree possible, that the various project elements coalesce to accomplish the project objective.

Matrix Project Management

In the matrix form of project management, a project manager with total responsibility for project completion is appointed. The project manager's responsibilities include planning, scheduling, acquisition and maintenance of all project resources, integration, testing, and implementation. While responsibility for the project rests with the project manager, all personnel assigned to the project remain in functional areas under the direct control of the functional managers.

The project manager's authority and responsibility for such a project flows horizontally across the organization. It is this apparent violation of more standard organizational theory that makes the matrix form of project management a difficult one to implement successfully. Although it is a workable and legitimate project form that has been successfully implemented in many organizations, its success depends on the presence of a number of conditions.

Matrix Project Management Problem Areas. One source of problems within the matrix form of organization is the apparent conflict between the authority of the project manager and the authority of the functional manager. Beyond the dimension of decisions discussed previously, the management and coordination of numerous project activities through the functional areas make the requirement for an effective project manager that much more important than in other project forms. As an example, a project manager within a matrix may have the authority to insist on thorough planning by the functional units as well as the freedom and authority to challenge the functional unit's project assumptions and the method in which work is performed. The exercise of this authority can cause organizational conflict between the project and the functional areas involved.

The solution to this problem rests with corporate management. Every member of the organization must be fully apprised of the project manager's role and the extent of his authority. Similarly, the role of the functional units that support the project must be defined in such a way that problems arising from authority conflicts can be avoided or reduced.

Another problem associated with the matrix form of project organization is the difficulty encountered in making project decisions. Because of the dual control inherent in this form, decisions affecting the project must be made with the awareness and concurrence of the functional areas involved. Frequently, however, the functional manager is faced with a number of priorities and decisions apart from the project requirement, making concurrence and commitment difficult to accommodate within the time frames required by the project manager. From the project manager's viewpoint, the involvement of the functional area in the decision process can be burdensome, prohibiting the ability to make timely decisions, with a resultant impact on project schedules and costs.

Other problems attendant with matrix project management are the increased requirement for coordination and integration of project tasks. The fact that tasks are dispersed throughout the organization gives rise to an additional level of interfaces that must be controlled by the project manager.

Matrix Project Management Advantages. The advantages of the matrix form of project management are that projects receive the level of attention required for successful completion with a minimum of disruption to the organization. Project members remain in their functional areas, reporting to the managers most familiar with how best to accomplish the tasks required.

This form further optimizes resource utilization in that project members, when not performing project tasks, can be assigned other tasks. Peer interchange and staff development are also facilitated.

Team Project Management

Perhaps the most easily recognized form of project organization is the team or task-force concept of accommodating project requirements. While the matrix form tends to be difficult to implement effectively and is only appropriate for large, highly complex projects (for which the project manager may not possess the technical skills to effectively manage tasks), the task-force approach is relatively easy to implement effectively and is best for medium-sized projects. In this form of project organization, the authority and resources required to complete the task are assigned to one manager, with the resultant effect that projects can be easily controlled and managed.

Project Team Problems. While this form of project organization would appear ideal because it eliminates many of the disadvantages of other forms, it is not without its problems. The apparent advantage of having complete authority over project personnel can have an alienating effect on those people assigned to the project. This arises because project members find themselves working for a manager who, upon project completion, will no longer influence their careers. In addition, sustained absence from their traditional functional organization can diminish their visibility throughout the course of the project.

Another problem is the absence of technical and professional interchange among peers. Task-force members frequently require interchange among people with the same professional background. Project commitments, however, frequently obviate the opportunity for this type of interchange.

Problems can arise within the functional areas as a result of the project team form. Functional managers are occasionally reluctant to assign their best people to a project team for fear that ongoing operations may be impaired. As a result, the people best qualified for the project may not be "available."

Furthermore, many line areas have procedural disciplines in place to facilitate operational efficiencies. These procedures might have to be compromised by the project manager for the benefit of the project. Project personnel returning to the functional areas, with the procedural disciplines in effect and in which other personnel may have been assigned their usual duties and responsibilities, can cause personnel problems and decreased functional effectiveness.

Project Team Advantages. There are a number of advantages to the project team form of project management. The fact that all resources are assigned directly to the project manager greatly facilitates directing and controlling the project. Project problems are more easily detected, enabling rapid remedial action. Decisions regarding the project can likewise be made and implemented quickly, as the functional areas need not necessarily be involved. Problems attendant with integrating the various project tasks are less severe in that direct control of the various project elements facilitates the integration process. Coordination is also less of a problem because of the centralized nature of the project team approach.

THE PROJECT MANAGER

All methods for project implementation except the functional approach require the appointment of a project manager, who has two basic functions: to develop a project plan and to implement it [3]. The formulation of a project plan depends, in large part, on the project manager's planning experience coupled with the use of numerous disciplines and tools for project scheduling and control. The implementation of the project, however, depends on the project manager's ability to focus the efforts of the project team on the resolution and completion of project tasks.

Project Manager Selection

Any current manager is a candidate for project management because the management skills of planning, organizing, staffing, controlling, and directing are necessary skills of the project manager as well. The difference lies in the emphasis of these management skills coupled with experience with project or task-force problems.

Formal education and experience in the management process are obviously indicated for effective project management. Education increases the manager's ability to learn from experience and enables creation of an environment conducive to effective management.

Education, however, is not the delineating factor for selection of a project manager. Experience, coupled with an awareness of the project management process and the ability to apply project management techniques, is required for the successful project manager [4].

Project manager selection is frequently difficult because of the lack of formal use of project management techniques. Within the construction and aerospace industries, where project management has become an integral part of the organization, the tendency is to develop project managers from the functional areas as experience on various projects is developed [5].

In selecting a project manager from the ranks of existing functional managers, emphasis should be placed on the manager's past performance on various assignments, together with how well that manager has evidenced the flexibility to adapt to different management techniques. The selection emphasis should be less on formal training in project management and more directed to the spectrum of experiences encountered by the prospective project manager. In general, the following attributes should be sought:

- A working knowledge of various fields of business and DP and the ability to delve into the intricacies of a specific technology
- A sound understanding of general management problems
- The ability to communicate effectively at all levels of the organization
- An integrative nature
- A strong background in planning and management

A survey of successful managers has indicated that the experiences that most contributed to their success were the opportunity for challenging assignments coupled with the opportunity to work for a variety of managers with varying styles [6]. It is this experience and opportunity that project management provides, with the additional potential benefit of creating more effective functional managers.

Project Manager Training

To date, very little in the way of formal training in project management has evolved from the academic community, although some schools are beginning to offer graduate-level courses directed at providing project management skills [7]. A number of seminars are available in project management, productivity management, and product management; some are specifically directed to DP personnel. These courses are beneficial for the first-time or experienced DP project manager and serve to augment the experience of managing a project. Despite the relative codification of project management as a management process, experience remains the best teacher of project managers.

The Role of the Project Manager

While the project and functional managers are similar in many ways, they differ significantly in their approaches to their jobs. Beyond the attributes of good management, the project manager differs from his functional counterpart in the following areas:

- Task orientation
- Management of personnel
- Management of interfaces
- Organizational emphasis
- Management perspective
- Implementation style

Task Orientation. Unlike a functional manager, the project manager is primarily involved with the planning and implementation of a one-time project. Because of this, he must be fully informed on the dimensions and limits of the project. The project manager must be sensitive to these limits to ensure that problems outside the scope of the project do not become a project requirement without full cognizance that the project, as originally defined, has changed. Within reasonable limits, the project manager's focus must be confined to the project requirements rather than redirected to the solution of problems within the functional areas with which he interfaces. Unlike his functional counterpart, he must continually "think project" and direct his energies to its successful completion.

Management of Personnel. In terms of personnel, the project manager's interpersonal skills and style tend to be different. Within a functional organization, personnel know fairly accurately their position with respect to salary, promotional opportunity, and personnel policies. When assigned to project tasks, these same personnel usually find themselves working for or being influenced by a manager outside of the functional organization. This divergence from traditional superior-subordinate relationships requires that the project manager use a modified style of motivation and persuasion to maintain the active support and participation of project members. To that end, the project manager must develop a project orientation among all project members and strive to maintain a project or team psychology.

Another potential problem with the project team approach is that the project manager is frequently unskilled in those areas represented by the project members. In addition, a project manager selected from one user department typically has little experience with the concerns of other user departments. This can result in the members feeling that they are technically mismanaged or that their problems and efforts are not fully understood or appreciated. To avoid this problem, the project manager must attempt to be thoroughly familiar, both conceptually and technically, with those functions required by the project.

Another source of personnel problems is the lack of visibility felt by team members as it relates to the functional manager. Sensitivity to this problem and continual communication with the functional manager regarding project member contributions can reduce this potential trouble area.

The matrix form also has the potential for personnel problems. The apparent violation of the superior-subordinate relationship gives rise to confusion and anxiety on the part of project members within functional areas. The project manager must again recognize this potential problem and modify his style to accommodate the matrix pattern. The project manager's emphasis must be on influencing the efforts of project members so as to reduce the potential for confusion and anxiety. He must be persuasive and effectively communicate throughout the various levels of the functional organization so that his relationship to all members of the project is clear.

Management of Interfaces. In the matrix form of project management, the emphasis is on the integration of project tasks and activities and the management of the interfaces created by the project organization [8]. As a project is broken down into tasks and subtasks and as these units of effort are performed by the specialized functional areas, the resulting matrix gives the project manager many more organizational and project interfaces to manage. These interfaces are generally personal, organizational, and systems interfaces [9].

Personal interfaces are those established among project members as well as those between the project members and the project manager. With members, the project manager may be called on to facilitate communications or resolve problems. Between the manager and team members, the project manager must strive to constantly improve the personal relationships and interfaces in order to maintain a harmonious environment.

Organizational interfaces exist between the project manager and his superiors as well as with the involved functional areas. Because each functional area may have its own objectives, disciplines, and functions, misunderstanding and conflict can easily occur at these organizational interfaces. The project manager must be aware of these potential conflicts and strive to effectively control these interface points.

CONCLUSION

Project management does not evolve by itself. What evolves is a trial-and-error process of project implementation that results in the ineffective management of change. Without effective project management, projects drift to completion (or go on and on) with, at best, only partial realization of the anticipated benefits.

There is no mystery to effective project management. It requires a sensitivity to organizational structures and a cognizance of the management problems inherent in managing a process that is different from the more traditional functions of the organization. Equally as important, it requires an approach or measured methodology to project implementation that ensures control of costs, time, and the quality of the completed project.

References

1. Stewart, John M. "Making Project Management Work." *Business Horizons.* Vol. 8 (Fall 1965).
2. Steiner, George A., and Ryan, William G. *Industrial Project Management.* Toronto: Collier MacMillan, 1968.
3. O'Brien, James J. "Project Management: An Overview." *Project Management Quarterly.* Vol. 8 (September 1977).
4. Cleland, David I. "Why Project Management?" *Business Horizons.* Vol. 7 (Winter 1964), p. 82.
5. Duke, Robert K., Wohlson, Frederick H., and Mitchel, Douglas R. "Project Management at Fluor Utah, Incorporated." *Project Management Quarterly.*
6. Patten, Thomas H., Jr. "Organizational Processes and the Development of Managers: Some Hypotheses." *Current Perspectives for Managing Organizations.* Edited by Bernard M. Bass and Samuel D. Deep. Englewood Cliffs NJ: Prentice-Hall, 1970.
7. Bernard, Prosper M. "Formal Training in Project Management." *1977 Proceedings.* Project Management Institute, Drexel Hill PA, 1977.
8. Lawrence, Paul R., and Lorsch, Jay W. "New Management Job: The Integrator." *Harvard Business Review.* Vol. 45 (November-December 1967), 145–49.
9. Stuckenbruck, Linn C. "Project Manager—The Systems Integrator." *Project Management Quarterly.* Vol. 9 (September 1978).

Bibliography

Gaddis, Paul O. "The Project Manager." *Harvard Business Review.* Vol. 37 (May-June 1959), p. 95.

Kast, Fremont E., and Rosenzweig, James E. *Organization and Management—A Systems Approach.* New York: McGraw Hill, 1970.

10 Structured Walkthroughs

by James A. Senn

INTRODUCTION

Before starting to construct any system, program, or module, it is advisable to ensure that the construction will be suitable. This is the purpose of a structured walkthrough. A structured walkthrough is simply a review of a system or a software product by people involved in or allied with the development effort; in other words, people at the same level in the hierarchy review a development effort together to find areas where improvements should be made.

Typically, structured walkthroughs are associated with programming and software construction. Such walkthroughs are aimed at uncovering errors in code. But this is only one way to use walkthroughs; they are also useful for design reviews. A team can review such design particulars as file types, access methods, data base design, planned coding schemes, and so on. Similar reviews can be undertaken at the requirements analysis stage so that any omissions, misunderstandings, poor decisions, or vague areas can be given attention before additional time and resources are committed to the effort.

The Need for Structured Review. Software systems are the lifeblood of the modern computer system. It is no secret that although hardware costs are decreasing rapidly, software-related expenses are rising at a rate that is totally out of control. Furthermore, even with the high cost of building or acquiring programs, it is difficult to obtain high-quality software. Even purchased software suffers from lack of quality. For example, the IBM OS operating system contained many errors; this was true even in later versions, which had been used at hundreds of field sites. This problem is not uncommon.

Although the industry has not yet devised a way to estimate software construction time accurately, it is clear that development time is often excessive. A goal of virtually all organizations is to obtain high-quality software more quickly, especially if this can be done at a lower cost than that currently incurred. Structured walkthroughs can help achieve this goal.

After software systems are built and installed, they are often used for many years, during which time they undergo changes. Some changes correct errors;

others add new functions or enhancements. In addition, maintenance is performed to rewrite sections of code so that a program is more efficient (see Table 10-1).

Unfortunately, even experienced programmers often write software that is not easily maintained. Possible modifications are not considered during the design and programming activities; thus, ease of maintenance is not designed into the code. By calling attention to this area of activity during design reviews, walkthroughs can aid in formulating design methods and coding standards that make program maintenance and enhancement easier. In addition, because proper use of walkthroughs can reduce the number of programming errors, the need for corrective maintenance can be reduced. Each of these advantages will become more apparent as this chapter examines ways to conduct software reviews.

Table 10-1. Maintenance Activities Associated with Software Systems

Maintenance Type	Description
Repair and Correction	To correct operating deficiencies, errors, and bugs that have been detected or are known to exist in the software or processing system or in the overall application design
Revision	To install mandated changes resulting from business or environmental (e.g., government) inputs that require modification of the system; also includes changes to improve job or program efficiency
Enhancements	Modifying an existing system or application to perform additional processing or reporting or to perform additional functions

The Purpose of Structured Walkthroughs. When conducting a walkthrough, the goal is simple: to find errors or problems. Note that no attempt is made to correct these difficulties at the time they are found. For reasons that will become more evident later, this is done sometime after the review is concluded.

It is also important to emphasize that structured walkthroughs are not intended to find fault with or blame any individuals. The design or the software, not the designers or programmers, is the focal point of the review. The emphasis of the discussion must be on improving the product rather than on assessing individuals. If this strategy is violated or abandoned, structured walkthroughs lose their meaning and value. In addition, involved personnel will begin to consider a required walkthrough an obstacle to be avoided rather than a helpful, cost-saving tool.

Relation to Training Programs. Training for people who are new to the department or new to a certain job is often neglected. Many organizations rely on some outside organization (e.g., a college or university or another business) or on attendance at professional seminars or workshops for training.

Although useful, they are not adequate because actual in-house methods and situations cannot be addressed.

A common way of introducing new programmers to a job environment is to assign them to a large-scale software maintenance project to clean up or adapt a software system to meet stated operation and quality specifications. Such projects are usually one to six months in duration. The rationale is that by correcting and/or enhancing someone else's work, new programmers can obtain skills and techniques while learning about common mistakes and how to avoid them.

Participating in structured walkthroughs extends such training methods. Walkthroughs can focus on live, in-house development or maintenance projects, at various stages of development (e.g., specification, design, and programming). Walkthroughs focus on problems and errors and also can show ways to detect and avoid these kinds of difficulties or inefficiencies.

Structured walkthroughs also offer training in a different sense. Each member of a walkthrough team must be an active participant, not a casual observer. While making contributions, each learns methods and gains insights from other people on the team. Thus, both junior and seasoned programmers learn and advance.

CONDUCTING STRUCTURED WALKTHROUGHS

The manner in which structured walkthroughs are conducted can largely determine their usefulness. In this section, the way to get the most out of this technique is discussed. The focus is on when to use the method, how to select appropriate participants, what procedures to follow, and what the result should be.

When to Do Structured Walkthroughs. Because walkthroughs are aimed at producing more reliable and cost-effective software products, there are several points in the software life cycle where walkthroughs can be applied: after the analysis stage, following design, and after programming. These reviews seem to be of equal value (see Table 10-2).

Requirements Review

Following investigation and determination of the information and processing requirements that the proposed design should satisfy, it is often useful to walk through the requirements specifications. This review, which is called either a requirements review or a specification review, is directed at examining the functions, activities, and processes to be handled by the forthcoming system. Any inconsistencies in the requirements stated by the users or identified by the analysts should be uncovered, as should any areas in the specifications that are vague or unclear. Inaccurate statements and assumptions should also be detected.

It is generally suggested that a requirements review focus on a written document that can be studied prior to the review session. Narrative descrip-

**Table 10-2. Types of Reviews Where Walkthrough Methodology
Can Be Used**

Type of Review	Description
Requirements Review (Specification Review)	Performed after a preliminary systems investigation has been completed in order to determine which functions and activities can or should be handled by the proposed application system. Also aimed at identifying misunderstandings, inaccuracies in specifications, or misleading assumptions on the part of either systems personnel or users.
Design Review	Performed to examine the logical design of the applications software. Assessment of the system blueprint as it has been established to determine whether the design will meet the original design specifications.
Code Review	Aimed at detecting problems in the coded software that stem from errors, misunderstandings, or poor adherence to programming standards. The review should ensure that the code meets the original design specifications. (Applies to both new developments and maintenance projects.)
Testing Reviews	Performed to ensure that the testing strategy being used for an application is sufficient to detect the most significant errors or bugs in an application program. Includes review of test data to be used.

tions that explain the context of the system are commonly used for requirements walkthroughs. These descriptions should spell out the different activities in the system area under investigation. Narrative descriptions should also identify key people and components and how they relate to one another. Sources and uses of information should also be identified.

Some shops use flowcharts that relate processes, control points, and data flow, or they use decision tables or even data dictionaries in place of narratives. The particular form of the description is unimportant, provided it is useful and understandable to those involved in the requirements review.

Design Review

As its name implies, a design review is a structured walkthrough to examine the logical design selected to deal with the information and processing requirements that were identified in the systems analysis stage. This review attempts to determine whether the proposed design is a valid one and whether it will meet the specifications. The walkthrough can be conducted by examining a design presented using any one of a number of documentation/presentation methods (e.g., HIPO and pseudocode).

Code Review

The type of review that is usually associated with structured walkthroughs is the code review. Most organizations and staff members begin using the walkthrough methodology with this form of review. Quite simply, a code walkthrough is an assessment of program code.

For new development projects, participants can review the entire software set as a complete package, comparing it with the original design or requirements specifications. Discovering that a portion of the code disagrees with the original specifications or finding problems or mistakes that originated with the earlier requirements analysis is not uncommon. Although having to modify the design or change the code at this point can be frustrating, it is easier and less costly than waiting until the software is installed.

As indicated earlier, code reviews should not be limited to new development projects; maintenance activities can also benefit from structured walkthroughs. The method is the same and the benefits equally significant. Maintenance projects have all of the attributes and problems of a new development activity, except that in maintenance projects some of the code and part of the documentation are already completed when the work starts (note that this can be an advantage or a disadvantage).

Testing Review

Many organizations overlook the advantages of using the walkthrough methodology to review testing, yet the same benefits of peer review can be realized at this life cycle stage, also. In fact, errors undetected during the testing stage will likely remain with the system when it is implemented, and it is these errors that cause nightmares for users and programmers alike.

Participants in testing reviews do not actually examine the output from test runs or search for errors that have been detected using a set of test data. Rather, they focus on the testing strategy to be used and determine whether it adequately detects critical errors. These people also assist in developing test data that can detect design or software errors. The purpose of testing is to find errors rather than to prove program correctness; therefore, an effective testing strategy is one that is likely to find the most serious errors.

THE PARTICIPANTS

One of the first questions to arise with structured walkthrough methodology concerns selection of participants. This issue and how it is handled can significantly affect the usefulness of the review strategy.

The Role of Programmers/Designers

Although programming and systems development have generally been considered solitary activities, this notion is changing. The concept of programming teams has become a topic of frequent discussion in many organizations. Through the team approach, more timely and reliable code is expected.

The use of structured walkthroughs is related to team programming, not just because they are based on similar objectives but because the walkthrough concept recognizes that systems and program development may involve several people for each step. Structured walkthroughs, then, are team-like activities. Furthermore, participants, with the possible exception of representatives from user departments, are individuals who are actually involved in developing software or applications.

The individual(s) who formulated the design specifications or wrote the code being reviewed as well as a number of other people, should participate in the walkthrough. Often a moderator is chosen to oversee the walkthrough and keep the group focused on the discussion topic (i.e., finding, not correcting, errors and problems). The moderator need not necessarily lead the review; many organizations prefer to have the programmer or developers do this because they are most familiar with the details of their project. (This familiarity, however, can be a problem by introducing strong biases or persuasive capabilities and causing reviewers to inadvertently overlook problems.)

It is imperative that the information produced during review sessions be captured completely and accurately. The leader of the session is occupied with ensuring that the appropriate concerns are discussed and therefore may not be able to jot down all of the points aired by participants. The programmer or developer may not record ideas in the same manner in which they are discussed by reviewers. Therefore, it is advisable to appoint a recorder for each walkthrough session in order to have all relevant details recorded completely and objectively. The intention is not, of course, to get a highly detailed record of who said what but rather to record the important points made. During the review, comments and suggestions may be made in rapid succession. Thus, the recorder must be constantly attentive. In many sessions, recorders are so busy taking notes that they cannot participate.

Experienced DP organizations are finding that standards for data names, module determination, field type and size, and so on are desirable. This is most often discussed in relation to data base environments, although it is equally important in non-data-base environments. In any case, the time to start enforcing these standards is at the design stage. They therefore can be discussed during the walkthrough sessions. This discussion can be led by the moderator or by a representative of the standards or data administration group.

Maintenance considerations should also be addressed during the structured walkthroughs; such concerns include coding standards, modularity, documentation, and parameterization. It is increasingly common to find organizations that will not accept new software for installation until it has been approved by software maintenance teams. In such an organization, a maintenance representative should be included in the review team.

Role of Management

Generally, management should not play a direct part in a walkthrough. Doing so would jeopardize the intent of the review. Reviews are aimed at helping individuals improve their design or product and, at the same time, creating a cost-effective software product for the organization. This discussion has characterized walkthroughs as occurring in an open, give-and-take atmosphere. Unfortunately, if management takes an active role in walkthroughs, it is likely that the true spirit of the review will dissolve.

Too often, management involvement is construed as evaluation. Many times, the result is that individuals attempt to perfect their product before the review session so that they look good in the eyes of management. Managers may feel that a considerable number of questions, mistakes, or changes indicates that the individual whose work is under review is incompetent. In brief, when management attends walkthrough sessions, the atmosphere changes significantly, resulting in less constructive results for the organization and the other participants.

Management may ask for reports summarizing the review sessions. Some types of reports, however, should not be produced. The only information that really need be passed on to management is that a review has been done, which project or product was discussed, and who attended or participated. Reports should not summarize the errors detected, modifications suggested, or revisions needed. If participants know this information is communicated, it will have the same effect as that of management actually observing. An appropriate sample evaluation summary is shown in Figure 10-1. This could be augmented with a 1- or 2-page memo giving a bit more detail.

Although it may seem unrealistic that management should not be involved in structured walkthroughs, most managers indicate that they prefer not to attend these sessions. They recognize that the walkthrough is a work session rather than a time to evaluate staff members. They also realize that because the sessions can be quite technical and clearly require a detailed knowledge of the product being reviewed, they would be unable to contribute much to the discussion. Moreover, managers are usually aware that their attendance can change the atmosphere of the sessions, thus inhibiting progress.

Size of Walkthrough Team

The members of the walkthrough team should be carefully selected so that the various roles are filled by competent and contributing people. Care must also be taken to ensure that the size of the team is appropriate for the project under review. At a minimum, the team should include the individual(s) who actually designed or coded the project, a recorder, and a leader.

In some organizations it is felt that having more persons involved in the examination increases the chances of locating problem areas. The group should not be so large, however, that lengthy discussions are needed; review sessions should not exceed 90 minutes. Considering the time constraint and general purpose of the review sessions, it is suggested that an upper limit of about seven persons be set for any walkthrough.

Summary of Walkthrough/Review

Date _____

Project/Contract No. _____   Time _____

Project Name _____

Unit/Section/Module Reviewed _____

Brief description of above _____

_____

_____

Participants

_____   _____
            Leader-Phone

_____   _____

_____   _____

_____   _____

Results

[ ] ACCEPT IN CURRENT FORM   [ ] REJECT — MAJOR REVISIONS
                                          NEEDED

[ ] ACCEPT WITH MINOR MODI-   [ ] REJECT — REDEVELOPMENT
    FICATION   NEEDED

[ ] REVIEW NOT COMPLETED

Discussion/Recommendations

_____

_____

_____

_____

_____

Attachments

_____   _____

_____   _____

_____   _____

                                    _____
                                          Leader Signature

Figure 10-1. Sample Form for Reporting Results of Walkthrough Session

Organizational Support and Participation

Although DP management should not have any direct role in walkthrough sessions, users should participate in such nontechnical walkthroughs as those conducted to examine specifications or functional requirements. Users can be extremely helpful in recognizing problems in system design attributes.

Some users may criticize walkthroughs as too time-consuming for the results they produce. This occurs because users do not fully understand the purpose and method of reviews or because they have had poor experience with structured reviews. In these cases the problem is not the review method but rather the way it has been implemented. In general, when structured walkthroughs are properly introduced and administered, the results are apparent to systems persons and users alike because more timely and correct systems are obtained.

PROCEDURES

Walkthroughs depend on fully informed participants; thus, those involved must come prepared. The individual requesting the walkthrough (i.e., the designer or programmer whose work is to be reviewed) should notify participants far enough in advance that they can study the materials to be examined. Generally, two to three days' notice is adequate.

Which materials should be distributed depends on the type of walkthrough. Copies of the documents or code to be walked through should be distributed, along with summaries of interviews, sample forms, and so on for a requirements review or system descriptions, I/O charts, and macro flowcharts for a design review. Code and test reviews usually require program listings and test data plans.

It has already been pointed out that walkthroughs should not be too lengthy. Because time and concentration limits preclude a single-session review of, for example, a 10,000-line COBOL program, it is obvious that this amount of code should not be distributed. Rather, only the modules actually being examined should be distributed. If this cannot be done easily, it may be an indication that the design is not sufficiently modularized.

It is essential that the participants in a walkthrough have the time, interest, and willingness to do the required preparation. If they cannot or will not prepare adequately, they will not be able to contribute to the walkthrough. It is better to have others replace them on the team so that maximum benefit can be realized. Note that if participants are expected to spend five or six hours in preparation, they may be justified in claiming they are too busy. In this case, it is probable that the session is not organized or limited properly. The objective of the session must be reformulated.

Starting the Walkthrough

There is more than one way to handle the mechanics of the walkthrough; the best approach depends on the organization, the nature of the people on the review team, and perhaps even the type of project being examined. It is generally suggested that the moderator for the session, rather than the programmer or designer, start the session and introduce the plan of action. The moderator may prefer to have the programmer or designer then give an overview of the project, presenting the important attributes of the design, code, and so on. It is recommended that the moderator ensure that this be an objective presentation that tells the what and how of the segment to be reviewed rather than the why. The presentation should not be a defense offered before the wolves start in with their cross-examination. If the review involves a project with which the participants are already familiar from previous walkthroughs, it may be unnecessary to begin with a formal introduction or overview.

How to Proceed

Depending on the type of review and whether or not it is the initial one, the actual walkthrough activities may vary. For first reviews of a project, attention should be given to determining what the logic or specifications are in comparison to what they were intended to be. For example, if processing logic does not perform all required validation checks, the team should discover this. As difficulties or misunderstandings are uncovered, they should be noted by the recorder so that they can be dealt with later.

If design specifications are being reviewed, participants should consider whether the proposed design will do the intended job and whether it will do it efficiently. Answering these questions necessitates knowledge of such items as file and transaction volumes, update frequencies, processing modes, access methods, keys, and the like. In addition, participants must know something about how the output from the system will actually be used; they should also be informed on the type of people who will use the system so that interface methods and protocols can be scrutinized.

When programs are reviewed, the participants also must be sensitive to execution efficiency, use of standard data names and modules, and program bugs. Appropriate comments and documentation permit this level of scrutiny. Obvious errors (e.g., syntax errors and blatant logic errors) can even be jotted down ahead of time by team members and submitted to the recorder, thereby saving meeting time. Other errors may merit discussion and examination during the review. Figure 10-2 shows a section of a checklist that might be used for noting problems and their severity.

If the review session is not the first one, there is "old business" to handle; that is, problems, suggestions, and comments mentioned during the previous walkthrough must be resolved. The designer or programmer should indicate how problems were solved, which suggestions were implemented, and which were not, along with the reason(s) for choosing alternative solutions. There may be very good reasons for not making changes as suggested, but they must be communicated to and agreed upon by other participants. When all old business is cleared, other areas can be reviewed.

Walkthrough/Review Checklist

Project/Contract No. _____
Project Name _____

Review Category	Absent	**Problem Detected** (check all that apply) Unnec	Error	Major*
Backup procedures				
Error messages				
Execution time				
External documentation				
Internal documentation				
Input validation				
Interface mechanism				
Procedural logic				
Passing of data				
Meets design specifications				
Meets user/problem specifications				
Meets coding standards				
Meets data standards				
Maintainability				
Storage use				
Test data				
Test procedure				
Test for end conditions				
Test for all possible conditions				
Test drop through				
Transfer of control				
Visible structure				

* Major error that will cause failure or crash

Figure 10-2. Sample Checklist for Guiding Review Activities

Approval

In many organizations, the team assigned to do structured walkthroughs on a particular project has final approval authority on the project. In other words, the team must approve the specifications, design, code, and test plan before the project can proceed to the next stage. In some cases, a project with problems that are corrected need not be returned to the team for a final review prior to its acceptance. When critical changes are involved, the team may decide that review of the modified program or design is required before acceptance is granted.

Some organizations have set up formal voting rules stating that unanimous agreement must be reached for acceptance. In others, a simple majority is necessary. In still others, the formality depends on the nature of the problem or change suggested; design questions may require full team approval, while questions of programming taste or execution efficiency may be settled with the approval of only a majority of team members.

PROBLEMS AND PITFALLS

As previously noted, problems can develop if management becomes directly involved in walkthrough work sessions or if participants do not adequately prepare for reviews. There are, however, other ways in which problems can develop.

If there is a tendency to try to complete a walkthrough too quickly (perhaps because the participants have some other meeting or project that is demanding their time), the walkthrough will not be fully successful; in the rush, problems or weaknesses in the product can easily be overlooked. It is generally better to postpone a walkthrough until the participants can devote the necessary time to it.

Some walkthrough teams get enmeshed in discussions of programming style. Although adherence to organization standards for such items as data names, field length, and type is important, inflexibility with regard to style can be counterproductive, especially if it is based on personal preference. Avoiding this type of difficulty is the task of the leader or moderator.

A similar concern regards individuals who feel they have *the* answer to a particular problem or situation. One of the objectives of structured walkthroughs is to set ego problems aside; however, an individual pushing a single correct approach is bound to occur. The participants and leader are responsible for recognizing and subduing such a person, even if it means openly ignoring his or her suggestions.

Egos are manifested in other ways, also. For example, some people enjoy making others look bad. They therefore attempt to find ''just one more error'' or discuss an approach as being ''the worst way to do it.'' Unfortunately, these problems are all too common.

Some pitfalls and problems are more subtle than the ones previously mentioned. The individual who is always late or the person who is in too many walkthroughs cannot do justice to any walkthrough; such a person needs to be better managed. Similarly, participants who fake it by making a leader or developer render extra detail as a way of masking their unpreparedness do not contribute to the team. Those who are unwilling or unable to face the real problems with a design, for example, and attempt to sidestep such problems by blaming the situation on standards or users represent a different problem.

CONCLUSION

Merely having a structured walkthrough for a program or project does not guarantee that the final product will be better than if no walkthrough had been held; the approach is effective only if used properly. To ensure success, the walkthrough must be properly managed and conducted in accordance with the guidelines outlined in this chapter.

Bibliography

Fagan, M.E. "Design and Code Inspections to Reduce Errors in Program Development." *IBM Systems Journal.* Vol. 15, No. 3 (1976) 182–211.

11 Post-Implementation System Review
by Jerome E. Dyba

INTRODUCTION

Systems development projects often fail to meet schedules, to conform to budgets, and to produce satisfied customers. In retrospect, we sometimes blame someone or something, but we seldom review the entire process and the results. What we should do is close the loop by performing a complete analysis of how we did what we did—after we did it.

A post-implementation review is designed to examine a development project and the resulting system to determine how effectively the feasibility study (including requirements and cost/benefit analyses) was performed, how completely development was consummated, how efficiently the computer operations staff is supporting the new system, and, most important, how satisfied the users are. A post-implementation review should provide an awareness of the achievements, shortcomings, and disappointments of the development effort and the system. This should then enable the organization to plan better and to improve the systems development methodology.

THE POST-IMPLEMENTATION REVIEW PROCESS

In performing a post-implementation review, it is important to determine:
- Whether the preliminary studies were complete
- Whether implementation progressed according to plan
- Whether the original cost/benefit analysis projections were accurate and what cost/benefit relationship exists today
- Whether the output, documentation, and security are adequate
- Whether the computer operations staff is able to meet the schedules and run the system successfully
- Whether any additional revisions or enhancements should be made to improve the system
- The value of the DP capability in assisting the organization to meet the daily service needs of its employees and recipients

The review culminates in a report that should tell DP management, user management, computer operations management, and development personnel how well implementation was effected.

When To Review

As a rule, the best time to perform a post-implementation review is approximately six months after system installation. During this period people can become familiar with the new system and can make minor corrections. This time also allows significant problems to surface. Earlier review does not allow costs and benefits to stabilize, nor does it allow time for people to relinquish old habits. Later review may have to deal with larger volumes, law changes, and the like, which tend to distort the scope and intensity of the original project. Naturally, some special consideration must be given to annual or other cyclical requirements.

If the report produced during the review is comprehensive, subsequent review is facilitated. Periodic reviews can be performed to ascertain whether changes are needed, whether the system should be overhauled, or whether a completely new direction should be considered.

Scope

When performing a post-implementation system review and analysis, all aspects of the current system should be reviewed, with emphasis on the users' point of view. Functions beyond the scope of the existing system (i.e., those that have not been systematized) should not be included in the study; however, they should be identified for possible future analysis.

Working closely with users, development personnel, and operations personnel is essential, especially for information gathering. A cooperative, open relationship must be developed to ensure a successful effort. During the study, minor modifications on which the reviewer and the primary user agree should be made. Any major modifications identified and agreed to should be considered a separate project(s) and scheduled according to other organization priorities.

All aspects covered in the Post-Implementation System Review Outline/ Checklist (see Appendix) should be reviewed, when applicable, in depth. The major sections in the outline are:
- General Evaluation
- Feasibility Study and Implementation
- Reports
- Data Base (or Master Files)
- Documentation
- Security
- Computer Operations
- Systems and Programming Maintenance

To reap the most benefit, the report that describes the findings of these analyses should be reviewed by management and followed up.

METHODOLOGY

The steps to be performed in a post-implementation review are:
1. Obtain management approval.

2. Inform users and DP personnel that the study is starting. (To maximize the findings, the reviewer should be ensured access to all levels of personnel in all pertinent areas.)
3. Have a kick-off meeting with the people involved.
   - Review the purpose of the study. Emphasize that this is not a witch hunt but a learning experience.
   - Review the outline of the areas to be covered.
   - Establish schedules and needs. (Two to six man-months are usually required.)
4. Obtain all information needed to review the system.
   - Interview user(s), DP personnel, and other involved personnel.
   - Use the Post-Implementation System Review Outline as a questionnaire, and obtain as much information as possible by observation and from involved personnel.
   - Visit other jurisdictions as required.
   - Research reports and the like as required.
5. Write a draft of the report, explaining in detail each item covered in the Post-Implementation System Review Outline. Review the draft with involved personnel and obtain sign-off.
6. Publish the final report. Produce a separate list of any recommendations.
7. Present the findings to management.
8. Follow up on implementation of the recommendations as they are approved.

CONCLUSION

Development projects have a beginning and an end; the post-implementation review is the end. It documents what was done, how successfully it was done, and what remains to be done. Anything after this review should be considered a new project and managed accordingly.

Incorporating a post-implementation system review into the normal development procedures helps solidify and improve the development process. The review closes the loop by accounting for all development project activities. The type of formalized and consistent approach described in this chapter and outlined in the Appendix should enable management to better plan, organize, direct, and control development projects.

APPENDIX

Post-Implementation System Review
Outline/Checklist

A. General Evaluation

The purpose of this section is to review the overall adequacy and acceptance of the system. User statements, explanations, and/or classifications should be fully described in the report.

1. General satisfaction with the system—This item is an interpretation of the users' experience with the implemented system. Comments should address:
   a. The level of user satisfaction
   b. The strengths of the system, areas of success, and so on
   c. Any problems and suggested improvements
   d. The extent to which the system is used (e.g., whether it is being worked around or used only as a last resort)
2. Current cost/benefit justification—This item documents whether the system is paying for itself. Details of costs and benefits should be provided in other sections; this section is intended merely to recap the costs and benefits. Comments should address:
   a. The extent of the benefits and whether they are less than or greater than the operating cost
   b. Whether the difference is permanent or will change over time
   c. Whether the system is or will be cost-justifiable
3. Needed changes or enhancements—This analysis gauges the magnitude of effort needed to improve the system. The report should contain the nature and priority of the suggested changes. Comments should address:
   a. The level of the required changes
   b. The suggested changes
   c. The extent of the required resystematization
4. Projected cost/benefit justification—This item projects whether future use of the system, after any needed or desired changes, will continue to be economical. Comments should address:
   a. The projected benefits and operating costs
   b. The extent of economic feasibility

B. Feasibility Study and Implementation

The purpose of this section is to gauge the completeness of the feasibility study and of implementation according to the study.

1. Objectives—This evaluation determines the adequacy of the original definition of objectives and whether they were achieved during implementation. An evaluation of whether the objectives have changed or should have changed should be included. Comments should address:
   a. The level of the objective definition
   b. The level of meeting objectives
   c. Possible changes to the objectives

2. Scope—This analysis determines whether proper limits were established in the feasibility study and whether they were maintained during implementation. The report should comment on:
   a. The adequacy of the scope definition
   b. The extent to which the scope was followed
   c. Possible changes to the scope
3. Benefits—This analysis determines whether the benefits anticipated in the feasibility study were realized. The report should detail all benefits, tangible or intangible, and any quantifiable resources associated with each. Comments should address:
   a. The adequacy of the benefit definition
   b. The level of benefits realized
   c. The anticipated benefits that can be realized
   d. The reason for the variance between planned and realized benefits, if any
4. Development cost—This analysis determines the adequacy of the development cost estimate and any deviation between the estimated and actual development costs. The report should address:
   a. The adequacy of the original and subsequent development cost estimates
   b. The actual development costs, by type
   c. The reasons for any difference between estimated and actual costs
5. Operating cost—This analysis determines the adequacy of the operating cost estimates and any deviation between the estimate and the actual operating costs. The report should summarize the resources required to operate the system. Comments should address:
   a. The adequacy of the operating estimates
   b. The actual operating costs
   c. The difference
6. Schedule—This evaluation determines whether implementation proceeded according to the predetermined schedule. The report should contain:
   a. An analysis of the scheduled implementation and actual conversion, including documentation, cut-over, training, and so on
   b. Specifics on the deviations from the schedule, if any, and the reasons for these deviations
   c. Identification of any speedups or delays
7. Training—This evaluation determines whether all levels of user training were adequate and timely. Comments should address:
   a. The timeliness of the training provided
   b. The adequacy of the training
   c. The appropriateness of the training
   d. Identification of training needs by job category
   e. The ability of the personnel to use the training provided

C. Reports

The purpose of this section is to evaluate the adequacy of and satisfaction with the outputs from the system. Care must be taken to ensure that all reports are evaluated. Comments about user capability to use the data provided are also appropriate.

1. Usefulness—This evaluation determines the user need for the output provided. The report should contain:
   a. Identification of the level of need as, for example:
      (1) Absolutely essential
      (2) Important and highly desirable
      (3) Interesting; proves what is already known
      (4) Unnecessary
   b. Demonstration of the ability to do without the reports
   c. Alternatives for obtaining the information
2. Layout—This analysis determines the layout aspects of readability, legibility, understandability, and the like. Comments naturally pertain to printed reports and screen formats. The following topics should be addressed:
   a. Date entries: as-of date, date prepared, for-period-ending date, and so on
   b. Headings: report name, columnar headings, unique report number, and so on
   c. Mnemonic expansion
   d. Totals
   Analysis of the report layout should also address:
   a. The understandability of the reports
   b. The degree of knowledge about each report that the user must have before making use of it
   c. Any problem areas
3. Timeliness—This analysis determines whether report production meets user needs. Comments should include:
   a. The frequency of output arriving on time, early, and late
   b. The amount of follow-up needed to obtain the reports
4. Controls—This evaluation determines the adequacy of the controls on master files or the data base, source documents, transactions, and outputs. Each area should be reviewed thoroughly for financial controls and file control counts. The report should address:
   a. The level of controls present in the entire system and on each component (e.g., transaction, batch, file)
   b. The adequacy of the controls; the strengths and possible areas for improvement
   c. The amount of resystematization required, if any
5. Audit trails—This analysis reviews the ability to trace transactions through the system and the tie-in of the system to itself. Comments should address:
   a. The thoroughness of the audit trails
   b. The level of improvements necessary, if any

D. Data Base (or Master Files)
The purpose of this section is to review the adequacy of the data base or master files. In analyzing a data base, some items may contradict each other, and these contradictions should be explained (e.g., completeness may be lacking while relevance is appropriate, or completeness may be high with relevance low).

1. Completeness—This evaluation determines whether the data base is all-inclusive and whether all needed or desirable data elements are included. The report should contain:
   a. An analysis of whether the data elements provided are:
      (1) Required
      (2) Desired
      (3) Required for future use
   b. The level of system supplementation with nonintegrated data that is required
2. Relevance—This evaluation determines whether the data base is too all-encompassing (i.e., whether there are data elements present that are never or seldom used). Comments should include:
   a. The frequency of data element use:
      (1) Frequently
      (2) Infrequently
      (3) Never
   b. Recommended changes
3. Currency—This evaluation determines the level of data element currency. The nature and use of the system dictate the need for currency. The system review report should specifically state the desired currency of data for meeting user/operational needs. The report should address:
   a. The desired currency of the data
   b. The currency achieved
4. Structure—This item evaluates the file structure used to ascertain whether other methods would be more appropriate. Alternatives could include:
   a. One long record for each entity
   b. Segmented records: a header plus numerous trailers
   c. Hierarchical data base structures
   d. Chained data records
5. Media—This analysis determines if data is on appropriate media or if others would be more appropriate. Alternatives could be:
   a. Punched card
   b. Magnetic tape
   c. Floppy disk
   d. Direct-access storage devices
   e. Mass storage devices
   f. Main memory
   Note: Analysis of media and/or structure may be more appropriately accomplished in a performance study that is independent of the post-implementation review.
6. Privacy (or allowed access to data)—This evaluation determines the adherence to restrictions on the access to data contained in the various files. The report should state desired privacy criteria for the system and then evaluate how they have been followed up to this point. The results should help to strengthen procedures in the future. Comments should address:
   a. The privacy criteria established
   b. Recommended privacy criteria
   c. Adherence to and violations of privacy

d.   The cost of providing this level of privacy
e.   The potential effect on individuals if the privacy criteria are not followed

E. Documentation

The purpose of this section is to review the adequacy of the published documentation and how well it has been maintained to date.

1.   Systems and user documentation—This review determines the adequacy of the overall documentation of the system. User documentation should be thoroughly appropriate for the user's purposes. The report should detail any weak aspects. The systems and user documentation should contain, at a minimum:
   a.   Systems narrative
   b.   Systems flowchart
   c.   Objectives, scope
   d.   Input and output documents (examples and explanations)
   e.   File specifications
   f.   Program narratives and flowcharts
   g.   Schedules for all jobs
   h.   Procedures for controlling the documentation
   i.   Security/privacy requirements
   The report should include:
   a.   A review of the completeness of the documentation
   b.   A statement about whether the documentation is up to date
   c.   The extent of any desired changes
   d.   The effort, if any, required to make the documentation comprehensive and current

2.   Operations run book—This review determines the status of the run books for control clerks and computer operators. At a minimum, the operations run book should contain:
   a.   A systems flowchart
   b.   Program history
   c.   JCL (jobstreams)
   d.   Labeling instructions
   Report comments should address:
   a.   The completeness of the run book
   b.   Whether the run book is current
   c.   The extent of suggested changes
   d.   The effort required to make the changes

3.   Data entry procedures—This evaluation assesses the adequacy of the data entry procedures. The report should review:
   a.   The completeness and currency of the procedures
   b.   The documentation for terminal users
   c.   The backup of formats and procedures
   d.   The extent of suggested changes
   e.   The effort required to implement these changes

4.   Program post lists—This item evaluates the filing and maintaining of post lists that correspond to the source decks (either in manual files or

on disk controlled by systems software). Comments should address:

   a. Completeness
   b. Availability
   c. Ease of locating the lists
   d. Currency
   e. Desired changes
   f. The effort required to make the changes

5. Test data and procedures—This item assesses the presence and maintenance of test data and the procedures for using it (to facilitate systems and program maintenance and to have predetermined data results for new equipment and software changes). The report should describe:

   a. The availability of the test data and procedures
   b. The currency of the test data and procedures
   c. The suggested changes
   d. The effort required to revise the test data and procedures

F. Security

The purpose of this section is to determine whether the system provides adequate security of files, data programs, and so on. In addition to access security, backup, recovery, and restart procedures should be reviewed.

1. Master data—This analysis determines whether adequate security, backup, recovery, and restart procedures are provided for master file data. The report should address:

   a. The adequacy of the security, backup, recovery, and restart procedures
   b. The suggested changes
   c. The effort required to make the changes

2. Transaction data—This analysis determines whether the security, backup, recovery, and restart capabilities adequately safeguard transaction data. Online systems naturally require special techniques (e.g., logging). The report should address:

   a. The adequacy of the security, backup, recovery, and restart procedures
   b. The suggested changes
   c. The effort required to make the changes

3. Source decks—This analysis determines whether the security, backup, recovery, and restart capabilities adequately safeguard the program source decks. The report should address:

   a. The adequacy of the security, backup, recovery, and restart procedures
   b. The suggested changes
   c. The effort required to make the changes

4. System-resident (SYSRES) pack—This analysis determines whether the security, backup, recovery, and restart procedures adequately safeguard the SYSRES pack. The report should address:

   a. The adequacy of the security, backup, recovery, and restart procedures

    b.   The suggested changes
    c.   The effort required to make the changes

5. Off-site storage—This analysis determines whether appropriate files, programs, and procedures are established to enable recovery from a disaster. The report should address:
    a.   The adequacy and currency of off-site storage procedures
    b.   The extent that procedures cover:
        (1) Master data
        (2) Transaction data
        (3) Source programs
        (4) Object programs
        (5) SYSRES pack
        (6) Documentation (e.g., systems, operations, user manuals)
    c.   The results of any adequacy-of-recovery test

G. Computer Operations

The purpose of this section is to ascertain the current level of operational activities. Although the user point of view should be primary, the computer operations view should also be investigated.

1. Control of work flow—This analysis evaluates the user interface with DP. The submittal of source material, the receipt of outputs, and any problems getting work in, through, and out of computer operations should be investigated. The report should address:
    a.   Any problems in getting the work accomplished
    b.   The frequency and extent of the problems
    c.   Suggested changes
    d.   The effort required to make the changes

2. Scheduling—This analysis determines the ability of computer operations to schedule according to user needs and to complete scheduled tasks. The report should address:
    a.   Any problems in getting the work accomplished
    b.   The frequency and extent of the problems
    c.   Suggested changes
    d.   The effort required to make the changes

3. Data entry—This analysis reviews the data entry function. The keying and data verification error rate is included in this analysis. Comments should address:
    a.   The volume of data processed (entry and verification)
    b.   The number of errors being made
    c.   The frequency of problems
    d.   The suggested changes
    e.   The effort required to make the changes

4. Computer processing—This analysis should uncover computer processing problems. Some areas to review are:
    a.   The correct use of forms, tapes, and the like
    b.   The ability of computer operators to follow instructions (e.g., forms lineup and proper responses on the console)

The report should address:
a. Identifiable problems
b. The extent of reruns, if any
c. A description of the work load
d. An evaluation of whether multiprogramming would be beneficial and, if so, how
5. Peak loads—This analysis assesses the ability of computer operations to handle peak loads and to clear up backlogs when they occur. Any off-loading that could be helpful should be investigated. Comments should address:
a. The level of user satisfaction
b. The adequacy of the response time (for online systems)
c. The effect of delays on online and/or batch systems
d. The suggested changes
e. The effort required to make the changes

H. Systems and Programming Maintenance

The purpose of this section is to evaluate the need for enhancements or revisions and/or the responsiveness to maintenance requests.
1. Systems maintenance—This review determines whether any changes should be made to the system to improve effectiveness or usability. Comments should include:
a. The suggested changes
b. The effort required to make the changes
c. Cost/benefit analysis of each
2. The volume of maintenance requests—This analysis determines the frequency and extent of maintenance requirements. The report should address:
a. The frequency of requests
b. The effort required to process the requests
3. Responsiveness—This analysis ascertains the level of responsiveness to user requests for systems and/or programming maintenance. The report should detail all requests that have been made, listing all open items. Comments should address:
a. The time required to accomplish each request
b. A follow-up of each satisfied request
4. Documentation maintenance—This investigation evaluates the currentness of the documentation in view of the maintenance requests that have been satisfied. The report should specifically address the status of the:
a. Systems and user documentation
b. Operations run book
c. Data entry procedures documentation
d. Program post lists documentation
e. Test data and procedures
The report should also contain the following information for all documentation:
a. When the last change was made
b. Plans for maintaining up-to-date documentation, if needed

12 Maintenance Documentation

by G.R. Eugenia Schneider

INTRODUCTION

Data processing is generally depicted in the literature as a world in which programs are carefully designed, written, tested, documented, used in production, and then replaced by more up-to-date programs created in the same way. The real world, of course, does not usually reflect this ideal. Programs and systems written many years ago are still in use; many of these originated as manual operations. When computers became available, operations that seemed amenable to automation were turned over to the machine, until the total system was eventually automated and the pieces were patched together with badly understood and thoroughly undocumented procedures.

This chapter primarily addresses the lack of documentation procedures for the particular needs of the maintenance shop. It discusses the functions of a maintenance shop and by whom these functions are performed, the procedures a maintenance programmer follows in updating a program, and the types of formal documentation a maintenance shop generates.

MAINTENANCE FUNCTIONS

A number of duties tend to be included under the banner of system maintenance; these duties can be categorized under four job titles:
- Maintenance manager
- Archivist
- Document librarian
- Maintenance programmer

Although one person is frequently expected to perform all of these duties, defining the jobs separately can help maintenance shops prepare for expansion.

Maintenance Manager

The maintenance manager monitors incoming program change requests and trouble-log entries. When a program change is proposed, the manager must arrange for a timely decision on the proposal's criticality and feasibility.

Sometimes systems changes are critical and must be made immediately. It is therefore especially important that changes be coded and tested, with no danger of the change destroying any previous version of a program or data entity. It is also important that the change can simply be withdrawn if it is ultimately disapproved. The manager must ensure that the requester has the opportunity to see the change in operation and that he or she gets a prompt response. Even if the change is withdrawn, the maintainers will gain valuable information about the program (possibly improving its documentation in the process).

The manager also assigns the writing of the formal documentation needed to support the maintenance effort and monitors the completion of documents. Although the librarian stores documents with other information about the system, program, and so on to which they refer, internal chronological document numbers should be assigned. These documents may include a system maintenance overview, a maintainer's guide, patch documentation, a formal definition of data files, format guidelines, and maintenance procedure guidelines.

The manager should review schedules and priorities monthly. This activity might well coincide with a general meeting of primary maintenance service users and maintenance staff. Such meetings provide an opportunity for a discussion of the overall software system goals and enable compromises between computer capabilities and project DP requests. The meetings also provide the best forum for a post-implementation review of software changes and can be used to ascertain that the changes meet the needs of, and are understood by, the users. Abbreviated minutes can be sent to management and may reveal a need for the allocation of additional resources for the maintenance function.

Status Report. The manager should submit a quarterly status report to his or her immediate supervisor. This formal documentation of the group's work provides much-needed publicity about the nontriviality of maintenance activities. The status report should contain the following information:

- Scope—the name of the reporting group and the period covered
- Work completed—a listing in outline form, by system and by program, of all significant code, procedure, and documentation changes made during the quarter
- Work scheduled—a listing of pending maintenance activities (comparing work scheduled this quarter with work actually required next quarter documents the unpredictability of assignments; it may help to justify additional resources)
- Reports acquired and generated—bibliographic references for new documents added to the document library during the quarter
- Personnel assignments—a list, by job title, of the man-months involved in these accomplishments

Among his or her many functions, the maintenance manager is frequently required to be a psychologist to keep users, customers, and programmers speaking to one another. It may also be wise to have the maintenance manager develop, implement, and maintain the organization's disaster plan. Few companies have worked out formal disaster plans, and the maintenance shop bears the brunt of the problem when the system crashes.

Archivist

The archivist is responsible for keeping up-to-date records of the contents of, changes made to, and backup copies of all computer tapes and files used by the maintenance staff. He or she develops and implements backup procedures for all files in the archive library. This includes all programs, utilities, data bases, data files, run streams, and the like that are stored on the computer and that have anything to do with programs currently being maintained. The archivist must generate run streams for listing the complete contents of any disk or tape file in the archive library. He or she must also know how to retrieve the latest version of any file as well as the previous versions that are under the jurisdiction of the maintenance shop.

The files in the archive library should be separated, first according to the computer on which they reside, then alphabetically by file (or library) name. Each file (or library) should have a documentary package in which listings are categorized as data, source code, run streams, or text. Within each category, individual physical elements should be stored in alphabetical order by the name used to access the element and then by revision date, with the most recent date first. The first section in the package should hold tables of contents of the file, again with the most recent entries listed first. Another section should contain listings showing how to retrieve that file from the archive tapes.

The archivist does not keep track of which files are logically associated with particular programs or with each other. His or her only concern is the maintenance and protection of the physical files and the keeping of records of the changes made to them. When any computer file is updated, therefore, it should be backed up immediately (on tape, card, or floppy disk) until the next archive tape is generated and verified. At least once a month, the file backup run streams should be modified to ensure that all files altered since the last formal backup are included in the next quarterly archive tape. At the same time, the table of contents of any altered file and listings of the new element should be stored in the file documentation package.

Before starting the archiving procedure, the archivist should request programmer confirmation of the completeness of the file change records assembled during the quarter. Then he or she generates the quarterly archive tape, documenting it with a list of the run stream that generated it. The tape is validated by listing the table of contents of each file or library on the tape, in order by tape file number. This output is then separated, and each table of contents is stored in the individual file or library documentation package. It should be noted that all files maintained in the archive library must be backed up at least once a year.

Document Librarian

The document librarian's primary function is to systematize the storage and retrieval of all information that is of use to the maintenance staff. Two other tasks, however, are routinely assigned to the document librarian. One is forms management, which involves storing master copies of all forms used in the shop and ensuring their copying and distribution. The other is WP control. Storing skeleton copies of typical documents generated by the maintenance group on the computer has dual advantages. It improves the motivation of documentation personnel by making it easier to produce the assigned documents and ensures that all documents of the same type and purpose have the same format.

The library should contain copies of all available information about the systems and programs being maintained. It is organized around a book of system abstracts, each referencing the programs, data files, procedures, and other entities that comprise the system. Library files are organized first by system, then by program and by data files. To avoid being buried under paper, the library should contain only one copy of any document. Documents defining data or interface procedures among programs in the system or relating to the design of the entire system are stored with the system documentation and are cross-indexed in the program documentation files. Similarly, documents defining individual programs or data files are stored under those headers and are referenced in the system documentation file.

When all information about a program is requested, the librarian should be able to determine the existence and location of such information immediately. A program that has been worked on will have library files, archive files, and a maintenance binder containing all of its documentation. If the program has not previously been worked on, the system should be sufficiently cross-indexed to show whether the program or any of its I/O files was referenced in the documentation of other programs.

If the requester is a programmer with a maintenance assignment, a library file and maintenance package for the program should be started. This package includes the appropriate forms for a set of basic documentation (called, for convenience, a minidoc)—a software abstract providing concise information about the software (see Table 12-1), a master run stream list showing how the program is run (see Table 12-2), and a run setup form, which is to be filled in by the user; a programmer's notebook for recording all pertinent information; a maintenance binder to hold the documentation; and copies of any existing documentation found in other library folders. If the search draws a blank, however, one should look elsewhere in the organization for "corporate memory" about the program. (A program that is just beginning its life in the maintenance shop requires the same paperwork.)

In developing a library system, it should be remembered that one of the major objectives is to indicate exactly what documentation exists concerning the software. The records are therefore categorized by type of documentation and by currency. New documentation should be added to the current documentation at least once a month, and the appropriate cross-references should be generated. New documentation files should be opened for programs and systems that have been assigned for maintenance for the first time or for which a first piece of documentation has been acquired; noncurrent documents should be transferred to a historical file. A list of all documents acquired and generated during the last quarter should be distributed at the end of that period.

Maintenance Programmer

The maintenance programmer seems to spend the day programming with the following cycle: make a change, make a run, curse, scribble, and loop. A few things need to be added, however. The programmer must record everything done, thought, or looked at that pertains to the program in his or her notebook—code changes and their effects, program runs and their outcomes, insights and information gleaned, useful conversations held, definitive progress made, milestones achieved, and the date on which the event occurred.

The programmer should insert any formal documentation or examples that might be used in such documentation into the maintenance binder, which typically includes:

- Minidoc—should be generated or updated by maintainer.
- User's guide—only if it exists; should not be written at this time.
- Analyst's manual—only if it exists; should not be written at this time.
- Maintenance information—patch or maintenance document; should be generated or updated by maintainer.
- Source lists and cross-references—should be inserted before maintenance work begins.
- Data files—formal definitions that should be generated or updated by maintainer.
- Sample I/O—should be inserted before maintenance begins.
- Benchmark I/O—should be present if it exists; should be generated by maintainer for permanent changes.
- System interactions—generally not formalized; should be included if interfaces with other programs become significant.
- Other—No file can be without this category.

All programmers' notebooks should be brought up-to-date at least weekly, preferably daily. If any changes have been made to data, program, or run stream files, the archivist should be informed and provided with a list of the new file contents. Newly acquired or generated documents should be sent to the document librarian monthly. By the last week of the quarter, the manager should be provided with an outline of progress made toward completion of each assigned project for inclusion in the quarterly status report.

MAINTENANCE GUIDELINES

The guidelines described in this section encompass most of the activities in a maintenance shop. These activities are divided into three major areas: an emergency takeover when a program that no one has ever heard of needs changing, permanent program changes, and temporary patches. The decision to make a temporary patch or a permanent modification rests with the maintenance manager.

Emergency Takeover

In an emergency situation, programmers must respond sensibly and without panic to the command, "Program XYZ doesn't work . . . fix it!" Assuming that no one in the shop has ever heard of XYZ, the programmer proceeds as follows:

1.   The librarian is asked for any information pertaining to XYZ. If any exists, the programmer should find out who worked on it last and should obtain the programmer's notebook and maintenance binder. If there is no information, the librarian should open a file and issue the paperwork for beginning a maintenance project.
2.   The appropriate forms used in a given shop for a programmer's notebook should be set up, and the programmer should immediately start to enter everything done, learned, and acquired about the program during the takeover period.
3.   If not already in hand from the archive, a table of contents of the program source file should be located, and listings (cross-referenced where applicable) should be generated of all program modules. If the archive has no information on this program file, a complete archive documentation package, including a table of contents and source listings, should be sent to the archivist, who assumes maintenance of these files.
4.   A complete set of sample I/O formats should be found or made, including dumps of any tapes or disk files used. If the programmer is fixing an aborted program, the failed run stream is needed as a test for the update; however, it is not sufficient as a benchmark run. For that, a data set that ran successfully before the failure is needed.
5.   A programmer's version of the change request should be written and sent to the requester (via the manager) for comment.
6.   Using whatever information can be gleaned from file lists, sample I/O, and the like, the programmer should work backward from the present to fill in the programmer's notebook. A new page should be allowed for each month to permit addition of later information.
7.   When the programmer's approved or amended change request comes back, the programmer should stop the book work and start the change—with a surprising amount of knowledge about the program.

Permanent Modifications

Permanent changes are generally those that fix a bug, add new capabilities, or improve usability. When designing a permanent change, it is important to remember that acceptance of the assignment is a commitment to the program in the eyes of the customer and the user. The next time a change is needed, the same programmer will be asked to make it; furthermore, he or she will be expected to come up to speed on the program and its intricacies almost instantaneously, even though several years may have elapsed since it was last seen. In self-defense, the programmer should design the change the way he or she wishes the program had been written in the first place. The programmer should consider how each type of change might be incorporated and should make current changes that will ease the insertion of future changes.

Code changes should be designed in accordance with structured programming techniques and such internal documentation concepts as those outlined in the next section of this chapter.

A test run stream that will change files only for the duration of a run should be used. All permanent file changes can then be made at one time, in order, when the change has been tested and is ready for implementation and production use. A hard-copy list of the final changes thus becomes a primary piece of documentation for the new program version.

A benchmark data set should be created for testing and validating the changes. The most desirable benchmark is representative of the full range of program use in general and of the proposed update in particular. It is crucial that the benchmark run successfully before modifications are made, or there will be no way to validate the update.

Before giving the finished change to production, one should ascertain that the previous version can be restored at any time. If the new version fails, the user should be given the familiar program immediately. It is not uncommon for the old version to crash harder than the new one because the error is not related to the recent changes.

Temporary Patches

Temporary program patches are made for a number of reasons, including:
- Special-purpose, limited-use modification
- Changes awaiting formal approval for implementation
- Quick-and-dirty changes to be added formally later

In general, it is unnecessary to know as much about a program to write a patch as to make a permanent change. The rationale for this is usually that although the change is needed immediately, it will be forgotten tomorrow.

Although the person who designs a patch is often much less closely identified with the program than the one who actually maintains it, patch changes tend to need defending more often. These changes are likely to be so out of phase with the original purpose of the program that it is hard to judge their total effect on program function. Defense will be demanded not only by customers and users but (especially) by the next programmer, who may have to incorporate the patch permanently and then live with it.

In designing a patch, the principles outlined in the section on Internal Documentation should be used. One additional caution—the entire change should be put in one unbroken block, in only one program module, if at all possible.

In implementing the patch, the only permanent file changes should be at the level of linked executable code; patch code should always be stored on the machine as an add or include file in the compilation procedure.

The test data required for patches tends to be somewhat informal. All that is needed is a data set that exercises the patch, whether or not it exercises the entire program. It is dangerous to assume that the program being patched truly works. If the test data does not work with the patched code, it is wise to see whether it functioned as planned on the original program.

A complete minidoc is written for the patched program before it is handed off to production. This task is relatively simple for a patch because the run setup needs to be revised only from the start of the run to the end of the altered code. Formal patch documentation (described in the next section) is rarely written before handoff because patches often carry unreasonable time demands. The programmer can then get additional information from user and customer responses before committing the activities to paper.

DOCUMENTATION FORMATS

Research in a large maintenance group at the Camp Pendleton Marine Base [1] has shown that the most important software documentation for a maintainer is the code itself. A far-off second is a narrative description of the purpose of each code module. Following that at fairly regular intervals are flowcharts, module hierarchy diagrams, data flow traces, and HIPO charts. Why then is other documentation required if well-organized lists are all a maintainer wants? One reason is that the code is probably not well organized. The major reason, however, is that if other documentation does exist, the maintainer must be able to find and use it. The following other types of formal maintenance documentation are therefore needed:

- Minidoc—used to find program records and indicate where the program is, who uses it, what it does, and where to find more information
- Maintainer's guide—used to formalize everything that a maintainer might want to know about a program
- Patch documentation—used to formalize a specialized change made to a program (from a maintainer's viewpoint) and to tell the customer what it is expected to do
- Data formats—used to define the physical and logical characteristics of data files and thereby provide information on interfaces among programs

Minidoc

The minidoc is basically the bare bones of information about a software item. It enables the librarian and archivist to file their records concerning the software, it informs customers and management of the software's capabilities, and it shows users how to set up runs with confidence. This sounds like a great deal of information, and it is; it also sounds like a great deal of paper, but it is not. The minidoc includes the software abstract, master run stream list, and run setup form.

Software Abstract. The software abstract (shown in Table 12-1) is intended to tell its reader enough about the software to indicate where to find all other pertinent information. A copy of the abstract opens every document concerning the software. Because it has a fairly rigid format and should be limited in length, a series of abstracts can be scanned swiftly, and the same information can always be found at the same place [2].

Master Run Stream List. The master run stream list (see Table 12-2) is, for the most part, an example of the way a program is run. The three types of entries in a master run stream list are the header, the control command list, and the data definitions.

Run Setup Form. The run setup form (RSU) is completed by the user whenever the program is run. It looks much like the master run stream, with a header that tells the program name, version, and the date it went into production; a line for each control command, with the unvarying parts of the command typed in all uppercase characters; and a line for each data type.

Commands. The command is printed with a space between each two characters, and the parts to be filled in by the user are represented as underscores. Information can be typed under the parts to be filled in, indicating the type of information needed and the column in which it begins.

Data Type Definitions. The definition of each data type is preceded by its type definition taken from the master run stream. If there can be only one occurrence of the data type in the run stream, the card is represented by a sequence of spaced underscores in the RSU. Under the location of each data item is the name of the item and an indication of those columns that belong to it. Elsewhere on the sheet is a user definition of the item, including such information as:

- Allowed values of coded inputs, and their meanings
- Type of data (e.g., binary, alphanumeric)
- Position in field (right or left justified or centered)

Unvarying data inputs are typed on the form. When there are several records of one data type, the RSU must include a table with enough lines for the maximum number of inputs. The table is preceded by the data definition from the master run stream. If the table entries have no intrinsic sequencing information, sequence numbers should be printed outside the table. The column headings on such a table should include the item name, the columns assigned to it, and the position code (e.g., R for right justified). A listing of user definitions of the data items in the table should appear elsewhere on the page.

Table 12-1. Checklist for a Software Abstract

Software Abstract

Name—software name; one-line description of purpose

Category—area of interest, department, system; superset name categorizing the software

Purpose—brief description of what the software does

Keywords*—quick-reference information further explaining software capabilities

Type—level of software (e.g., program, system, processor, subroutine library, utility, data base)

Date—relevant life-cycle dates (e.g., ordered, designed, implemented, revised)

Status—two-part description of software status: experimental or certified; fully or partially supported (and by whom)

Documentation—bibliographic references for formal software documentation

Contacts—names, locations, and phone numbers of cognizant people (those who can answer questions about the software)

Comments—any additional brief information that tells a reader whether the software might be interesting and useful

Abilities/limitations*—brief description of parameters and restrictions on use

Source*—where software was obtained and for what computer it was written

Language*—computer (or data base) language in which software is written

Access*—where source code, data files, and run streams are stored; commands needed to obtain such information from the computer

Testing*—reference to any documentation on testing or indication of how exhaustively the software has been validated

Date/initials of preparer

* Optional items; to be included if space and motivation permit

---

Maintainer's Guide

The maintainer's guide is the maintenance-shop equivalent of an analyst's manual in the design shop and is meant to be a complete standalone definition of a program. This guide is often the only existing formal documentation of the program, and the format reflects that possibility. The table of contents has an interest code to the right of each entry, indicating who would be interested in that section. The usual codes are: u—user, e—project engineer, and p—programmer; many more are possible. A maintainer's guide is outlined in Table 12-3.

Patch Documentation

Although patch documentation has the same format as the maintainer's guide, it makes no attempt to define the entire program. Only changes made since the last production version are defined, whether or not any previous documentation was written about the program at any level.

The author should be aware that this document (like the maintainer's guide) may well be the only formal description of the program. Every effort is

Table 12-2. Contents of a Master Runstream List

Header Record—one line, containing:
'**' . . . MASTER RUNSTREAM'
Program Name (version)
Cataloging date
' . . . **'

Example:

MASTER DECK—GLIST (VER. 3)—CATALOGED 15 AUG 77

Control Card Definitions—one line for each control command in a complete program run, listed in the order in which they would appear in a typical runstream. All unvarying command elements are typed uppercase, exactly as they would be punched. Varying command elements are typed lowercase, if that option is available, or designated by a character that would not otherwise appear in a command line. If there is room, an explanation appears to the right of the command, telling what information should be filled in.

Example:

@ASG, T 14,U9,XXXX—FLIGHT TAPE NUMBER REPLACES X'S

Data Definitions—For any data type that is part of the runstream (i.e., not tape or disk files, just card input), there is one line of definition, located where that data type would appear in the setup. The data definition contains:
'Dn.'—sequence number (first card is D1.)
Name of the data type
'('
Number of cards of this type
'REQUIRED' or 'OPTIONAL'
If optional, the condition under which the data element is used
')'

Example:

D3. SEGMENT CARD (IF3 CARDS—OPTIONAL—IF IF3.GT.0)

---

therefore made to point out, by the format of the document and the audience flags in the table of contents, which parts of the patch documentation are to be extracted for users, analysts, or project management. The outline for patch documentation is shown in Table 12-4.

Data Formats

A general format for documenting data files is shown in Table 12-5. There are, of course, some slight differences involved for effective documentation of special sorts of data files. The difference in defining a card file, for example, would be the columns occupied by each data item and possibly a notation that the item is right or left justified in its field.

Sequential Files. Tape files require such additional information as the number of files, the packing density, and a complete definition of any check sum or block start and end codes added by the system. If the tape is to go to another organization, the recording speed, make, and model of the tape unit

Table 12-3. Outline for a Maintainer's Guide

Table of Contents

Cover Letter—memo explaining the purpose of the document, the authority under which the work was performed, and the distribution of the memo and the document

Abstract—a copy of the program abstract from the minidoc

Description of Program Use   u
   Operating Instructions
   Control Commands
   Input Data Cards
   Other Input Media
      Overview
      Disk files
      Magnetic tape
      Other

Description of Program Outputs   e,u
   Output-Handling Procedures
   Printer Outputs
   Other Output Media
      Overview
      Plotter
      Disk files
      Magnetic tape
      Other

Definition of Program Structure   p
   Summary of Physical and Logical Design
      Functional Description
      Module Flowchart
      Data Flowchart
      Other Design Information
   Main Program
   Description of User Subprograms
      Name and function
      Inputs—arguments, globals, and read-ins
      Outputs—arguments, globals, and write-outs
      Functional description—purpose, algorithms, and so on

Appendices

Appendix A—Sample Inputs and Outputs   e,u
   Benchmark runstream listing
   Input data for benchmark run
   Printer outputs
   Other outputs

Appendix B—Special User Information   u
   Master runstream
   Run setup form
   Other

Appendix C—Definition of Data Files*   p
   Input files
   Scratch files
   Output files

Appendix D—Special Maintenance Information   p
   Runstream listings
      Catalog production program
      Document contents of program files
      Document contents of data files
      Compile and test
      Other
   Machine-dependent information

Table 12-3. (cont)

System subroutines and libraries used
Internal data representations (and character codes)
Time and memory parameters
Other

Appendix E—Source Lists and Cross-References   p
  All program modules

---

* See Table 12-5.

Table 12-4. Outline for Patch Documentation

Table of Contents

Cover Letter—Memo explaining the purpose of the document, the authority
under which the work was performed, and the distribution of the memo
and the document

Abstract—A special program abstract that describes the program as it operates with the patch in place

Changes in Program Use   u
  Operating Instructions
  Control Commands
  Input Data Cards
  Other Input Media
    Overview
    Disk files
    Magnetic tape
    Other

Changes in Program Outputs   e,u
  Output-Handling Procedures
  Printer Outputs
  Other Output Media
    Overview
    Plotter
    Disk files
    Magnetic tape
    Other

Changes in Program Structure   p
  Summary of Physical and Logical Changes
    Functional Description
    Module Flowchart
    Data Flowchart
    Other Design Information
  Procedural Changes in Existing Routines
    Main Programs
    Subprograms
  Description of New Subprograms
    Name and function
    Inputs—arguments, globals, and read-ins
    Outputs—arguments, globals, and write-outs
    Functional description—purpose, algorithms, and the like

Appendices

Appendix A—Sample Inputs and Outputs   e,u
  Test runstream listing
  Input data for test run
  Printer outputs
  Other outputs

Table 12-4. (cont)

Appendix B—Special User Information   u
   Master runstream
   Run setup form
   Other

Appendix C—Definition of Data Files*   p
   New or altered input files
   New or altered scratch files
   New or altered output files

Appendix D—Special Maintenance Information   p
   Runstream listings
      Catalog patch version of program
      Document altered contents of program files
      Document altered contents of data files
      Compile and test
      Other
   Machine-dependent information
      System subroutines and libraries used
      Internal data representations (and character codes)
      Time and memory parameters
      Other
   File updates
      Program updates
      Data updates
      Runstream updates

Appendix E—Source Lists and Cross-References   p
   New or altered modules only

---

* See Table 12-5.

---

used to write the tape should be included. For disk files, especially if the file will be read by a language or processor unlike the one that wrote it, physical track size limits should be included, as well as any check sum and start- or end-of-block information the system generates.

Random-Access Files. The additional information needed for random-access files includes all possible values of the key variable and the exact command used to access a record. The listing appended to the documentation should be sorted by the most common order of the keys.

Data Bases. There is usually a straightforward way to list the logical format description for data bases. This listing, and the commands used to produce it, should be included in the documentation. The file list should contain all current contents in an order that reflects the most common use.

Internal Documentation

Except in rare programs written according to structured design precepts, the code to be modified may be all but unreadable. It is also possible that no external documentation exists for the program and that the only information about the program's inner workings is the code itself.

Table 12-5. General Format for Documenting Data Files

Cover Page	Identification
	Source (organization and computer)
	Type (e.g., card, tape, disk)
	Cognizant personnel (location, phone, and area of cognizance)
	End-file definition (physical and/or logical)
	Block definition
	Length—records per block or variable
	Frequency—generally, when a block is written
	Format—what record types are included, how many of each, and in what order
	Recognition—how start of block is recognized
	Record definition
	Length—physical and logical record size (words, characters, bytes, or bits)
	Frequency—as above, what causes a record to be written
	Format—detailed physical and logical description of the contents of each data item
	Word definition
	Length—in bits
	Complementation—ones or twos
	Format—bit-by-bit description of internal format of all data types in the record
	Processing history
	Source program(s) where used
	Description of transformations within program
	Parameters and restrictions on content and use
Data Item Table	
	Sequence number
	Logical definition (with units)
	Range of values or list of possible values
	Scaling information (i.e., units conversions)
	Coding (e.g., ASCII, 029 keypunch, BCD)
	Format (e.g., nnn.nn, F6.2, real)
	Data name in program (at time of I/O)
	Comments
Listings and Dumps	

The following paragraphs provide guidance on how to structure maintenance updates and their supporting comments to make the update as readable as possible and the program more maintainable.

Module Structure. Before attempting to modify a piece of code, it is wise to understand the current structure of that code and to evaluate the feasibility of restructuring the module to enhance its readability and maintainability. A maintenance programmer who has the opportunity to design a new module or restructure an old one should keep in mind such desirable features of well-structured code as:

- Dividing code into logical blocks, with transfers of control into the block only to the first statement in the block

- Making all transfers of control downward, except for iteration (loop) control, and always to a label on a line that contains no executable code
- Putting all data definitions in one location and initializing all working data values
- Whenever possible, making all transfers of control to points outside the block from the last statement of the block
- Including debug printouts of representative inputs and outputs of the module and for any intermediate steps where the data has been significantly transformed
- Segregating into discrete modules the sections of code that implement main control flow, machine- or installation-dependent code, input, output, and standardized procedures

Main Program Header. The header, or preface, in the main program should tell a reader enough to enable him or her to control the program, run a benchmark, and understand the results. At a minimum, the header should contain the information shown in Table 12-6.

Subprogram Header. All independent modules in the program, whether or not the source code is stored separately from the main program, should have a header block. A minimal subprogram header block should contain the information outlined in Table 12-7.

Embedded Comments. Generally, the only comments within the code should be the logical block separators. These should be three lines long, with the first and third lines merely a comment character and a row of asterisks in columns 30 through 72. The middle line should contain a few beginning and ending asterisks (e.g., columns 30 to 33 and 70 to 72), a logical block ID number, and a few words describing the purpose of the block.

If the compiler allows it and if the program may be compiled on another machine, in-line comments should be used to explain all the data declarations. It must be remembered that not all information is contained in comments; a great deal can be imparted by judicious use of:

- Mnemonic variable and constant names
- Defining constant values in data declarations and setting initial variable values with executable code
- Indenting the beginning of a line to make the logical and iterative structure as obvious as possible
- Writing I/O formats with the variable names, column headings, and I/O format specifications lined up vertically in the source list

WHEN IS THE JOB FINISHED?

Two frequently asked but often difficult to answer questions are: When is a maintenance task finished? When is a newly developed program ready for production? The primary difference in the answers is the level of documenta-

Table 12-6. Contents of a Header for a Main Program

Title—one-line descriptive name of program

Author—name, address, and phone number of author

Date Written/Implemented—date when the program was authorized for creation or date of first production use (whichever is more significant)

Function—paragraph that defines the purpose of the program, the source of the inputs, the destination of the outputs, and any significant logical transformations in between that may not be readily obvious from reading the main program code

Revision Table—table showing, for each (known) version of the program:
Revision number or letter code
Date the new version went into production
Name or initials of the update programmer
Change request code (to identify the programmer's notebook and maintenance log entries for the revision)
One-line description of the reason for the revision

Subroutine and Function Table—table that describes, for each separate code module: type code (e.g., u—user, l—library), name, one-line description of purpose

Peripheral Requirements Table—table that describes, for each peripheral device used:
Type code (i—input, o—output, s—scratch)
Logical unit name and/or number in the program
One-line description of use
Source or destination of the data
Exact control-card image (if any) used to assign the peripheral device to the run

References—bibliographic references to documentary information about the program, including background references for algorithms or procedures, and manual references documenting unusual programming techniques

Parameters and Restrictions*—any situations in which data may be inappropriate for the use of this program

Formal Runstreams*—name, location, and purpose of canned runstreams to aid in program maintenance and use

* Optional

---

tion that can reasonably be generated before a program is handed off to production. Following the guidelines presented in this chapter greatly simplifies the question of when to hand off to production. Very briefly stated, a program is ready to be given back to its users when:

1. The test run using the benchmark data has been approved by the programmer and the customer
2. The minidoc has been updated to reflect the change, and the user understands the new run setup form

The maintenance task is a slightly different matter; the programmer must also:

1. Bring the maintenance binder up to date
2. Complete and close the programmer's notebook
3. See that the maintenance or patch documentation is written, approved, and sent for distribution
4. Give all appropriate materials to the librarian and archivist to bring their files up to date

Table 12-7. Contents of a Header for a Subprogram

Access—information about where to find the module on a computer file (includes version identification and date of most recent modification)

Function—a few lines describing the purpose of the module

Inputs—name and short definition of each datum that is used within, but whose value is set outside, the module

Outputs—name and short definition of each datum whose value is set within, for use later outside of, the module

Revision Table*—as shown in Table 12-6.

Data/Parameter Definitions*—name and definition of significant data entities; define any data held in global storage at this point

Equations*

Parameters and Restrictions*

References*—documents describing algorithms, procedures, or unusual programming practices used in the module

* Optional

When the librarian and archivist have also closed their files and the post-implementation review has passed without new changes being required, the maintenance project is finished.

CONCLUSION

Statistics indicate that maintenance may represent up to 80 percent of resource time and cost during the lifetime of a software system [3]. This knowledge, unfortunately, is rarely translated into adequate maintenance facilities and resources. Part of the reason is that no one really understands what software maintenance entails nor how many people it takes to start a maintenance shop. A rule of thumb is that at least two people are needed at the outset: a programmer and someone in a general support function. Where the shop goes from there depends on a variety of factors, including:

- Size of system(s) being maintained
- Complexity of interactions among programs
- How often changes are requested
- How soon changes must be finished
- Whether a design shop exists to trade new programs for old

Probably the most important factor of all, however, is how well the code was initially written.

How can a system be designed for maintainability? This chapter has attempted to answer that question. If an organization practices good structured programming techniques, backed up by solid structured documentation practices, and adds a formalized test and evaluation procedure for enhancements and updates, systems will be more maintainable from the start.

References

1. Shumate, Kenneth C., and Anderson, Gordon E. "Resolving DP Management Issues by the Numbers." *Data Management*, Vol. 18, No. 2 (March 1980), 32–35.
2. Schneider, Gene, French, Don, and Lucas, Lee. "How to Document Software." Naval Weapons Center CCF-87. China Lake CA, August 1977.
3. Lientz, B.P., Swanson, E.B., and Tompkins, G.E. "Characteristics of Application Software Maintenance." *Communications of the ACM*, Vol. 21, No. 6 (June 1978), 466–471.